Understanding Elections through Statistics

Elections are random events.

From individuals deciding whether to vote, to individuals deciding who to vote for, to election authorities deciding what to count, the outcomes of competitive democratic elections are rarely known until election day... or beyond. *Understanding Elections through Statistics* explores this random phenomenon from three primary points of view: predicting the election outcome using opinion polls, testing the election outcome using government-reported data, and exploring election data to better understand the people.

Written for those with only a brief introduction to statistics, this book takes you on a statistical journey from how polls are taken to how they can—and should—be used to estimate current popular opinion. Once an understanding of the election process is built, we turn toward testing elections for evidence of unfairness. While holding elections has become the *de facto* proof of government legitimacy, those electoral processes may hide the dirty little secret of the government, illicitly ensuring a favorable election outcome.

This book includes these features designed to make your statistical journey more enjoyable:
- Vignettes of elections, including maps, starting each chapter to motivate the material
- In-chapter cues to help one avoid the heavy math—or to focus on it
- End-of-chapter problems designed to review and extend what was covered in the chapter
- Many opportunities to turn the power of the R Statistical Environment to the enclosed election data files, as well as to those you find interesting

The second edition improves upon this and includes:
- A rewrite of several chapters to make the underlying concepts more clear
- A chapter dedicated to confidence intervals, what they mean, and what they do not
- Additional experiments to help you better understand the statistics of elections
- A new introduction to polling, its terms, its processes, and its ethics

From these features, it is clear that the audience for this book is quite diverse. It provides the statistics and mathematics for those interested in statistics and mathematics, but it also provides detours for those who just want a good read and a deeper understanding of elections.

Ole J. Forsberg, PhD, is an Associate Professor of Mathematics-Statistics and the Associate Dean of Faculty Affairs at Knox College in Galesburg, IL. He received a PhD in Political Science from the University of Tennessee-Knoxville in 2006, concentrating in International Relations, War, and Terrorism. After finishing his dissertation, Forsberg began a deeper investigation of the statistical techniques he used there. As a result of the ensuing embarrassment, he began formal graduate statistical studies at the Johns Hopkins University (MSE, 2010) and concluded them with a PhD in Statistics from Oklahoma State University in 2014 (Go Pokes!). His statistics dissertation explored and applied statistical techniques to testing elections for violations of the "free and fair" democratic claim.

His research agenda lies in extending and applying statistical methods to elections with the goal of drawing as much out of the available information as possible... whether this is in terms of understanding political polls, predicting election outcomes, or testing elections for unfairness.

Chapman & Hall/CRC
Statistics in the Social and Behavioral Sciences Series

Series Editors
Jeff Gill, Steven Heeringa, Wim J. van der Linden, and Tom Snijders

Recently Published Titles

Introduction to R for Social Scientists: A Tidy Programming Approach
Ryan Kennedy and Philip Waggoner

Linear Regression Models: Applications in R
John P. Hoffman

Mixed-Mode Surveys: Design and Analysis
Jan van den Brakel, Bart Buelens, Madelon Cremers, Annemieke Luiten, Vivian Meertens, Barry Schouten, and Rachel Vis-Visschers

Applied Regularization Methods for the Social Sciences
Holmes Finch

An Introduction to the Rasch Model with Examples in R
Rudolf Debelak, Carolin Stobl, and Matthew D. Zeigenfuse

Regression Analysis in R: A Comprehensive View for the Social Sciences
Jocelyn H. Bolin

Intensive Longitudinal Analysis of Human Processes
Kathleen M. Gates, Sy-Min Chow, and Peter C. M. Molenaar

Applied Regression Modeling: Bayesian and Frequentist Analysis of Categorical and Limited Response Variables with R and Stan
Jun Xu

The Psychometrics of Standard Setting: Connecting Policy and Test Scores
Mark Reckase

Crime Mapping and Spatial Data Analysis using R
Juanjo Medina and Reka Solymosi

Computational Aspects of Psychometric Methods: With R
Patricia Martinková and Adéla Hladká

Principles of Psychological Assessment
With Applied Examples in R
Isaac T. Petersen

Multilevel Modeling Using R, Third Edition
W. Holmes Finch, Jocelyn E. Bolin, and Ken Kelley

Understanding Elections through Statistics: Polling, Prediction, and Testing, Second Edition
Ole J. Forsberg

For more information about this series, please visit https://www.routledge.com/Chapman-
-HallCRC-Statistics-in-the-Social-and-Behavioral-Sciences/book-series/CHSTSOBESCI

Understanding Elections through Statistics
Polling, Prediction, and Testing
Second Edition

Ole J. Forsberg

CRC Press is an imprint of the
Taylor & Francis Group, an **informa** business

A CHAPMAN & HALL BOOK

Second edition published 2025
by CRC Press
2385 Executive Center Drive, Suite 320, Boca Raton, FL 33431, U.S.A.

and by CRC Press
4 Park Square, Milton Park, Abingdon, Oxon, OX14 4RN

CRC Press is an imprint of Taylor & Francis Group, LLC

© 2025 Ole J. Forsberg

First edition published by CRC Press 2021

Reasonable efforts have been made to publish reliable data and information, but the author and publisher cannot assume responsibility for the validity of all materials or the consequences of their use. The authors and publishers have attempted to trace the copyright holders of all material reproduced in this publication and apologize to copyright holders if permission to publish in this form has not been obtained. If any copyright material has not been acknowledged, please write and let us know so we may rectify in any future reprint.

Except as permitted under U.S. Copyright Law, no part of this book may be reprinted, reproduced, transmitted, or utilized in any form by any electronic, mechanical, or other means, now known or hereafter invented, including photocopying, microfilming, and recording, or in any information storage or retrieval system, without written permission from the publishers.

For permission to photocopy or use material electronically from this work, access www.copyright.com or contact the Copyright Clearance Center, Inc. (CCC), 222 Rosewood Drive, Danvers, MA 01923, 978-750-8400. For works that are not available on CCC, please contact mpkbookspermissions@tandf.co.uk

Trademark notice: Product or corporate names may be trademarks or registered trademarks and are used only for identification and explanation without intent to infringe.

ISBN: 978-1-032-68453-6 (hbk)
ISBN: 978-1-032-62186-9 (pbk)
ISBN: 978-1-032-68455-0 (ebk)

DOI: 10.1201/9781032684550

Typeset in Latin Modern
by KnowledgeWorks Global Ltd.

Publisher's note: This book has been prepared from camera-ready copy provided by the authors.

As with everything I do, I dedicate this book to my family and friends. Without them, I would not be here today.

Contents

Preface to the Second Edition — xiii

Acknowledgments — xix

List of Figures — xxi

List of Tables — xxv

About the Author — xxvii

Symbols — xxix

1 **Introduction to Polling** — 1
  1.1 The Importance of Political Polling — 3
  1.2 Polling Terms — 5
  1.3 Sampling Methods — 6
  1.4 Biases in Polling: The First Words — 10
  1.5 Legitimacy in Polling — 12
  1.6 Experiments in Polling — 14
  1.7 Conclusion and Looking Ahead — 17
  1.8 Extensions — 18
  1.9 Chapter Appendix — 19

2 **Simple Random Sampling** — 21
  2.1 Simple Random Sampling (SRS) — 24
  2.2 One Estimator of π: The Sample Proportion — 24
  2.3 SRS without Replacement — 35
  2.4 Conclusion — 36

2.5	Extensions	37
2.6	Chapter Appendix	38

3 Interval Estimation 43

3.1	Margin of Error	45
3.2	Confidence Intervals	50
3.3	Credible Intervals	55
3.4	Conclusion	61
3.5	Extensions	62
3.6	Chapter Appendix	63

4 Stratified Sampling 68

4.1	Stratified Sampling	71
4.2	The Mathematics of Estimating π	74
4.3	Confidence Intervals	85
4.4	Conclusion	89
4.5	Extensions	90
4.6	Chapter Appendix	91

5 The Bayesian Solution 98

5.1	Overview: Why Bayesian Analysis	100
5.2	Bayes Law	103
5.3	A Particularly Helpful Shortcut	108
5.4	The Bayesian Calculations	110
5.5	Bayesian p-Values	114
5.6	Conclusion: Current and Future Use in Polling	115
5.7	Extensions	116
5.8	Chapter Appendix	117

6 Aggregating Polls 121

6.1	Overview	123
6.2	Simple Averaging of Polls	124
6.3	Weighted Averaging of Polls	126
6.4	Averaging of Polls over Time	128

	6.5	Looking Ahead	134
	6.6	South Korean 2017 Presidential Election	138
	6.7	Conclusion	142
	6.8	Extensions	143
	6.9	Chapter Appendix	144
7	**Going Beyond the Two-Race**		**151**
	7.1	Overview	152
	7.2	The Multiple Comparisons Problem	153
	7.3	The *Real* Problem	156
	7.4	The Bayesian Solution	158
	7.5	A Brief Return to Fisher	164
	7.6	Conclusion	166
	7.7	Extensions	167
	7.8	Chapter Appendix	168
8	**Testing Elections: Digit Tests**		**170**
	8.1	Electoral Forensics	172
	8.2	History and Insight	173
	8.3	The Benford Test	175
	8.4	Extending Newcomb and Benford	180
	8.5	Using the Generalized Benford Distribution	183
	8.6	Conclusion	190
	8.7	Extensions	191
	8.8	Chapter Appendix	192
9	**Testing Elections: Regression Tests**		**201**
	9.1	Differential Invalidation	203
	9.2	Regression Modeling	210
	9.3	Examining Côte d'Ivoire	220
	9.4	Conclusion	222
	9.5	Extensions	222
	9.6	Chapter Appendix	223

10 Polling Analysis: Brexit Vote, 2016 — 229
- 10.1 Know Your Data — 230
- 10.2 Combining the Polls — 236
- 10.3 Discussion: What Went Wrong? — 237
- 10.4 Conclusion — 239

11 Election Analysis: Collier County — 241
- 11.1 The CVR Files — 242
- 11.2 Collier County Background — 244
- 11.3 Fundamental Question — 248
- 11.4 Data Preparation and Exploration — 250
- 11.5 Conclusion — 256

12 System Analysis: Sri Lanka Since 1994 — 257
- 12.1 Differential Invalidation — 259
- 12.2 Methods and Data — 259
- 12.3 Results by Election — 262
- 12.4 Discussion — 273
- 12.5 Conclusion — 274
- 12.6 A Sri Lankan Postscript — 275

13 An Afterword: Biases in Polling — 276
- 13.1 The Sources of Bias and Variance — 277
- 13.2 A Concluding Diatribe: The Blame for 2020 — 281

A A Brief Introduction to R — 282
- A.1 Installing R on Your Computer — 283
- A.2 A Quick, Sample Session — 284
- A.3 A Second Example — 291
- A.4 Conclusion — 294

B Four Probability Distributions — 295

- B.1 A Brief Introduction to Probability … 295
- B.2 The Binomial Distribution … 297
- B.3 The Multinomial Distribution … 308
- B.4 The Beta Distribution … 309
- B.5 The Dirichlet Distribution … 311
- B.6 End of Appendix Materials … 315

Bibliography — 317

Index — 331

Preface to the Second Edition

Elections hold a special place in my heart. As a Political Scientist, they are the outward manifestation of the hopes, dreams, and aspirations of a people toward their government, their society, and their futures. As a Statistician, they are a random process producing reams of data that should be describable, predictable, and testable.

In both cases, elections tend to fall short of those ideals. They tend to be expressions of our fears instead of our hopes. They tend to be heavily influenced, both indirectly in the form of the media — both social and not — and directly in the form of the governments that hold the elections, count the ballots, and report the results. As Nicaraguan leader Anastasio Somoza stated in a 1977 interview with the London Guardian,[Eme]

> Indeed, you won the elections, but I won the count.

Elections *are* random variables, but they are random variables without known (or knowable?) distributions, except in the simplest of cases with the strongest of assumptions. This makes testing elections for direct government intervention difficult, to say the least.

And yet, here I am writing a book dedicated to the proposition that elections can be statistically understood, with that understanding giving us a deeper insight into ourselves and where we want our future.

For Whom

To achieve my goals in the first edition of this book, I decided on as broad an audience as I could. I missed the mark by quite a bit. This second edition takes the lessons I learned from that experience, as well as some new trends in polling, and improves upon my original idea.

To be *absolutely clear*, this is not a graduate-level textbook. My audience remains those who have had some exposure to statistics. This exposure could be from an advanced high school course or an introductory college course in statistics. It may also come from extensive experience in employment, such as what professional journalists covering elections and polls would obtain.

Thus, I envision this book to be accessible to those who have already had an introduction to statistical thinking in terms of the ideas behind hypothesis testing. This means having exposure to confidence intervals, test statistics, and p-values.

The point of the book is to obtain a better understanding of elections, from the polls to the predictions to the testing. This requires some mathematics. However, the amount of mathematics required depends on the goals of the reader. The more you want to understand polling, the more mathematics you must learn and use. However, one can obtain a sound understanding of polls and elections without the full mathematical treatment.

In short, I try to make the mathematics self-contained and relegated to the chapter appendices. Doing this allows you to skip over it without losing the fundamental points of the book. However, working through the mathematics will aid in understanding the assumptions behind elections.

The Plan

This book is divided into two main parts: modeling and testing. The **first part** explores modeling elections. This requires understanding polling and how polls relate to each other. This part introduces Bayesian analysis as a complement to the usual frequentist (or "Fisherian") statistics we all learned in our introductory statistics course.

The first chapter provides an introduction to polling and its terminology. It covers some general sources of bias introduced through the polling process. This is bias we understand. It is bias that we, as ethical members of the profession, try to minimize. It is also bias that we can model once we understand it. Thus, I also introduce the idea of experimentation in statistics.

Chapter 2 introduces us to statistical concepts by way of the simplest type of legitimate polling scheme—simple random sampling. It is the simplest because its assumptions make the mathematics much simpler to follow. Because of this, the reader develops two understandings: the relationship between the sampling scheme and the mathematics, and the meanings of the estimation intervals (both confidence and credibility).

The third chapter focuses on confidence and credible intervals. These provide the necessary context to *any* report on poll results. Claiming the candidate has 53% support is both misleading and meaningless without also including the precision of that estimate. In this chapter, we explore margins of error, confidence intervals, and Bayesian credible intervals. These all quantify the uncertainty in the estimate, thus making it useful.

The next chapter begins with explaining the usual sampling method used in polling — stratified sampling — and how that choice affects the mathematics behind the estimation. It continues with complicating the sampling scheme to include controlling for some of the randomness in the individual polls by using auxiliary information like phone type, gender, and political party membership. At the end, one should better understand how polling is actually done and the importance of several of the assumptions made.

Chapter 5 gives an introduction to Bayesian analysis in the realm of political polling. There is a philosophical chasm between Fisherian and Bayesian analyses. The former is what we are used to from our intro-stats course. The latter offers a cleaner paradigm that also allows us to use past information and to make probability statements about the support in the population for a position or candidate. I believe that Bayesian analysis is the future of polling analysis, so this is (perhaps) the most important of the mathematically inclined chapters.

The sixth chapter looks at combining polls. Thus far, we have examined each poll in a vacuum. If two polls separated by time give different results, then we tend to conclude that the views of the population have changed. However, each poll result is a random variable. Polls separated by time *will* differ, even if the population view has not changed. In this chapter, we take this realization and explore how to combine polls... and how doing so affects estimation intervals.

The **second part** concerns testing election results for evidence of unfairness. This part is divided into two chapters based on the amount of information available for testing. In each case, the tests are for violations of the "free and fair" claim made by democratic governments. Such violations may arise from some systemic unfairness in society, some systematic unfairness in a particular election, or some outright fraud committed by the government or its supporters.

Chapter 8 deals with testing for fraud and unfairness when the only available data is the vote counts in each electoral division. This is the most difficult case to test because we must rely on the distribution of the digits in the counts. The two distributions covered are the Benford distribution and the generalized Benford distribution. In neither case is it clear that it is the *correct* distribution. I conclude the chapter regretting the existence of the Benford test, for it is misused in election testing.

In the ninth chapter, we learn how to properly use the invalidation rate to test for fraud and unfairness. Much fraud and unfairness arise because ballots are invalidated for certain subgroups at a much higher rate than for others. This differential invalidation gives rise to certain patterns that may be detected using statistical methods.

Next are **three chapters** that provide in-depth examples of election analyses. The first looks at the Brexit polling and explores what "went wrong." The second is of Collier County, Florida (home to Naples). Here, we can see the data available and the story it tells. The last is a formal test of persistent electoral unfairness in Sri Lanka.

Finally, Chapter 13 serves as the conclusion to the book. It looks more at biases in polls and what we can all do to mitigate their effect. I offer a call-to-action for all of us to take ownership of the polls and understand their strengths and limitations.

Why R

From what I can tell, the vast majority of books that use R spend a lot of time explaining why they use R. For me, choosing R comes down to three main reasons: it is free, it forces the analyst to be explicit, and it is extensible. The R Statistical Environment costs nothing. Because it is a scripting language, one can simply read the scripts to see the analysis steps. Finally, if you want R to do something that it cannot (easily) do at the moment, one can either write a function or use one of the many, many, many packages available — for free.

I use multiple R versions, depending on the computer I am using. R tends to put out a new version twice each year. However, all of the analyses done in this book work as well on my R3.0.1 (Windows XP laptop) as on my R3.0.3 (Windows Vista desktop) or my R4.3.0 (Windows 10 desktop) or my R4.2.0 (macOS laptop) — as long as one doesn't count the speed differences caused by the underlying computer speed.

One can download the most recent version from the Comprehensive R Archive Network, affectionately known as "CRAN," from this URL:

https://cran.r-project.org/

To help with learning the basics of using R, CRAN offers many manuals. In addition to this, a simple Internet search will return a large number of sites dedicated to helping you learn how to use R.[R 23]

I do provide a quick "How-To" in R as Appendix A, starting on page 282. If you have no experience with R, I strongly urge you to work through that appendix to get started.

A Very Special Acknowledgment

At the start of every chapter, you will find at least one map. That is for two reasons. The first is that I find maps infinitely interesting. The stories they tell are clearly connected to the underlying distribution of people. The second reason is that they look cool. I realize this is not the most academic reason, but it is definitely true. From my conversations with students, I have discovered that maps draw them into the story much more efficiently than a typical scatter plot.

Thus, it is with great honor that I explicitly thank GADM, the Database of Global Administrative Areas, for providing the shapefiles from which I created the maps.[GAD20, GAD24] Not only does GADM have the shapefiles for all of the countries in the world, allowing one to create their own maps, but they have many pre-created maps such as average annual temperature, elevation, and total annual precipitation.

And so, thank you!

Conclusion

And so, it is now time to implement the above plan. Turn the page (and a few more beyond that) and start your journey to a deeper understanding of the statistics of elections around the world.

$\sim$ Ole J. Forsberg
Galesburg, IL, USA
Spring 2024

Acknowledgments

I would like to express my gratitude to all who have helped me get to where I am. This includes Knox College for providing me with the support necessary for writing a manuscript of this length. It also includes the members of the Junior Faculty Research group that gave valuable feedback on the first edition of this book.

I would also like to acknowledge the many students and researchers I have had the privilege of working with over the past quarter-century. They have all helped to shape this text in some form. This especially includes Elliot Bainbridge, who worked his way through the unpolished version of the first edition, and Nea Schramm, who worked her way through the second edition. Both gave me insight into what worked, what didn't work, and what still needed to be done. It also includes an entire class that used this book as the text for my JOUR 195: Interpreting Political Polls course. That class rocked!!!

Finally, I would like to thank Daniel Naiman for starting me in the right direction, Walter Mebane, Jr., for laying a strong foundation in electoral forensics, and Mark Payton for pushing whenever I started to slack off.

As always, the errors are mine and the perfections are theirs.

List of Figures

1.1	Maps of the 2020 Taiwanese Presidential election	3
1.2	Illustration of the three populations	6
1.3	Distribution of sample proportions	16
2.1	Maps of the 2014 Scottish independence referendum	23
2.2	MSE graphic for the Agresti-Coull estimator	31
2.3	Histogram of sample poll results	34
3.1	Maps of the 2024 Bhutanese general election	44
3.2	Where the 95% comes from	50
3.3	Confidence interval illustration	66
4.1	Map of 2016 US presidential election results	70
4.2	Effects of weights on stratified sampling	78
5.1	Map of 2023 Turkish results for the YSGP.	99
5.2	Comparison of prior and posteriors	107
5.3	Comparison of prior and Özdemir posterior	113
6.1	Map of 2017 South Korean presidential election results	122
6.2	Graphic of Hong poll numbers in the final month.	129
6.3	Graphic illustrating rectangle weighting.	131
6.4	Three possible weighting functions.	132
6.5	Graphic illustrating triangle weighting.	133
6.6	Hong poll numbers with final election result	135
6.7	Hong poll numbers, with the WLS estimation line.	137
6.8	Moon poll numbers	139

6.9	Ahn poll numbers	141
7.1	The outcome of the 2016 presidential election in Gambia	153
7.2	Multiple comparisons problem	155
7.3	Support for the three Gambian candidates	157
7.4	Posterior support using a non-informative prior	160
7.5	Posterior support using an informative prior	164
7.6	Correlation between support levels	165
8.1	Map of 2009 Afghan presidential election results	171
8.2	Page from a table of logarithms	174
8.3	Initial digit distribution for AFG 2009	178
8.4	Extended Benford distribution	182
8.5	Estimated log-likelihood distributions, OK and AFG	187
8.6	Estimated log-likelihood distributions, LIT and CIV	188
8.7	The Log-UNIF$(0,6)$ probability density function	194
8.8	Benford BENF$(0,\mathbb{N})$ distribution	197
9.1	Maps of 2010 Ivoirian presidential election results	202
9.2	Maps of 2010 Ivoirian presidential invalidation rates	204
9.3	Invalidation rate signature	206
9.4	Differential invalidation effects	209
9.5	Comparing four modeling methods	219
9.6	Comparing four Ivoirian models	221
10.1	Map of 2016 Brexit results by country	231
10.2	All Brexit-2016 polls	232
10.3	Brexit-2016 polls by mode	233
10.4	Brexit-2016 polls by population	234
10.5	Brexit-2016 polls showing support level	235
10.6	Estimated Brexit support	236
11.1	Collier County voting pyramids	246
11.2	Voting history of Collier County	247

List of Figures　　　　　　　　　　　　　　　　　　　　　　　　xxiii

11.3	Histogram of ballots cast	252
11.4	Vote counting path	253
11.5	Support levels in each of the five regions	255
12.1	Map of 2015 Sri Lankan presidential election results.	258
12.2	Invalidation plot for the 1994 presidential election	265
12.3	Invalidation plot for the 1999 presidential election	266
12.4	Invalidation plot for the 2005 presidential election	268
12.5	Invalidation plot for the 2010 presidential election	269
12.6	Invalidation plot for the 2015 presidential election	270
12.7	Invalidation plot for the 2019 presidential election	271
12.8	Map of 2019 Sri Lankan presidential election results.	272
12.9	Invalidation curves for all six elections	273
13.1	Map of the 2016 and 2020 US Presidential Elections	278
A.1	The R console window after starting R.	285
A.2	The R window after tiling the two sub-windows.	286
A.3	Results of preliminary analysis of the UNIF$(0,1)$ dataset.	290
A.4	Results of preliminary analysis of positioningtube dataset.	293
B.1	Probability functions for a Binomial distribution	298
B.2	Comparison of Normal and Binomial	300
B.3	The pmf of the example's distribution.	301
B.4	Confidence interval for a two-person race.	304
B.5	Display of a Three-Race for the example.	305
B.6	Display of a Three-Race for the example.	306
B.7	Display of a Three-Race.	307
B.8	The pdf of three Beta distributions.	310
B.9	Graphic of St. Ives outcomes	314

List of Tables

2.1	The Binomial and the Hypergeometric distributions.	36
4.1	Weights used by Monmouth	69
4.2	Monmouth cross tabulation summary	88
6.1	Estimates for Hong based on weighting function	134
6.2	Estimates for Moon based on weighting function	140
8.1	Partial counts from AFG 2009	177
8.2	The p-values for the 2009 Afghan Presidential election	179
8.3	Multinomial averaging examples	190
11.1	Demographics of Collier County voters	245
12.1	Regression results for the six elections	263

About the Author

Ole J. Forsberg, PhD, is an Associate Professor of Mathematics-Statistics and the Associate Dean of Faculty Affairs at Knox College in Galesburg, IL. He received a PhD in Political Science at the University of Tennessee-Knoxville in 2006, concentrating in International Relations, War, and Terrorism.

After finishing his dissertation, Forsberg began a deeper investigation of the statistical techniques he used there. As a result of the ensuing embarrassment, he began formal graduate statistical studies at the Johns Hopkins University (MSE, 2010) and concluded them with a PhD in Statistics from Oklahoma State University in 2014 (Go Pokes!). His statistics dissertation explored and applied statistical techniques to testing elections for violations of the "free and fair" democratic claim.

His research agenda lies in extending and applying statistical methods to elections with the goal of drawing as much out of the available information as possible... whether this is in terms of understanding political polls, predicting election outcomes, or testing elections for unfairness.

Symbols

Symbol Description

α The claimed Type I error rate. A Type I error occurs when one rejects a true null hypothesis. The usual goal of hypothesis testing and estimation is to ensure that the actual Type I error rate is close to α.

β The claimed Type II error rate. A Type II error occurs when one rejects a true alternative hypothesis.

$\mathcal{L}$ The likelihood of observing the data. It is calculated as the product of the individual probability functions.

μ The *population* mean (expected value). This is the long-run average of experimental outcomes.

π The *population* proportion. This is usually interpreted as the proportion of the voters who will vote for a specific position, candidate, or party.

p The *sample* proportion. This is the proportion in the particular sample collected.

$\pm$ "plus-minus." This symbol indicates that the equation will have two results. The first result uses the $+$; the second, $-$.

$\propto$ "is proportional to." This is used to focus only on the variable of interest and sweep aside the normalizing constants.

σ The population standard deviation, the square root of the population variance.

θ A generic symbol for a population parameter. This could represent μ or π or σ or some other population parameter of interest.

Θ A generic symbol for the possible values of the population parameter (a.k.a. the parameter space). For example: if the population parameter is π, then Θ is values between 0 and 1; $\Theta = [0, 1]$. If working with a specific parameter, the capital Greek letter is used. Thus: $\Pi = [0, 1]$.

Z_p The pth quantile in the standard Normal distribution. By convention, the absolute value is used. Usually, one will see $Z_{\alpha/2}$. This is the $\alpha/2$th quantile.

xxix

1

Introduction to Polling

Every great journey requires a first step. This is yours. It is a step toward understanding elections through statistics. It is a step toward understanding many of the moving pieces in conducting a legitimate poll. It is a first step toward understanding the strengths and limitations of statistics and reality.

This chapter serves as an introduction to the many aspects of polling, from conducting them to analyzing them to reporting them. As a professional, we need to ensure that what we do is ethical, that we report reality instead of trying to affect it. I will be returning to this theme throughout the book.

Republic of China: The Presidential Election of 2024
Each chapter of this book starts by introducing some election that I find interesting. The purpose of this is three-fold. First, it emphasizes that polling and elections are of worldwide interest. Second, it provides some concrete events to help frame the material of the chapter. Finally, I just love talking about elections and how they affect the world around us.

For this first chapter, let us look at the Taiwanese presidential election of 2024. Note that wading into another country's affairs is always fraught with danger. This is especially true when there is a frozen conflict regarding territory.

Different countries may claim sovereignty over the same chunk of land (and its people). Conflicts over these types of claims tend to be rather long-running, outliving those living at the time of onset. Some examples include territorial conflicts over Kashmir, Palestine, and Tibet.

A fourth example is the conflict over China. Both the People's Republic of China (PRC) and the Republic of China (ROC) claim sovereignty over both mainland China and the islands off its east coast. However, the PRC only controls mainland China, and the ROC only controls Taiwan.[1]

The Republic of China (Taiwan) sits off the coast of mainland China. Since Chiang Kai-Shek and the Kuomintang (KMT) were expelled from the mainland in 1949 by the Communists, Taiwan has been held by the ROC but coveted by the PRC.

If not for the importance of both countries, this conflict may already have ended. However, it has not. In fact, it is becoming more important in the 21st century, with the wealth of the ROC and the growing wealth of the PRC. This conflict is even more touchy with Taiwan having a rather successful political party supporting independence and with the PRC seeking to unite all China under its flag.

As with most countries, one important difference between the major parties is economic. The Democratic Progressive Party (DPP) are economically center-left, while the Kuomintang (KMT) are center-right. However, given its history and its relationship with the People's Republic of China, there is another axis controlling the ROC parties: independence vs. unification. Without question, the PRC is pushing for unification per its one-China policy. Thus, this political axis is intimately tied to the PRC and its positions (both politically and geographically). Taiwan's two main parties have subtly different positions along this axis.

According to its charter, the Democratic Progressive Party (DPP) is pro-independence, with the following clarification:

> Based on the fundamental rights of the people, the establishment of a sovereign Taiwan Republic and the formation of a new constitution shall be determined by all citizens of Taiwan through a national referendum.[Wu01]

The Kuomintang (KMT), on the other hand, is less in favor of independence for Taiwan and more in favor of the one-China policy as the Republic of China). With that being said, the KMT strongly opposes the PRC's "one country, two systems" option that was imposed upon both Hong Kong and Macao.[EK22]

Into this milieu, we have the 2024 presidential election in the Republic of China. Because the incumbent Ing-wen Tsai won in both 2016 and 2020, she could not run for another term; that is, she was **term-limited** as president.

---

[1] Please note that there are other territories involved in this conflict. For instance, both the PRC and the ROC claim Tibet, and the ROC claims "Outer Mongolia." The author, who is suddenly speaking in third-person, has his own ideas about what territory belongs to whom. He wishes to appear neutral in this conflict, focusing instead of the global importance of the ROC.

The Importance of Political Polling

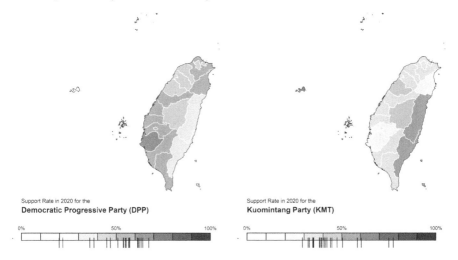

FIGURE 1.1
Result maps of the 2020 presidential election in the Republic of China (Taiwan). The left map illustrates the proportion of votes supporting the Democratic People's Party (DPP); the right, the Kuomintang (KMT). These data are from Taiwan's Central Election Commission.[Cen20]

This left Ching-te Lai as the candidate from the DPP. Against him were Yu-ih Hou of the KMT and Wen-je Ko of the third-party Taiwan People's Party (TPP).

On November 24, 2023, the candidates had to officially register. On that day, ETtoday conducted a poll of 1348 people.[ETt23] Of those surveyed, 34.8% said they supported Lai of the DPP, 32.5% supported Hou of the KMT, and 21.2% supported Ko of the TPP.

How should we interpret these polling results, especially in light of how the two political parties performed in the 2020 election (see Figure 1.1)? More importantly, how can this information be used to help the three parties elect their candidate?

1.1 The Importance of Political Polling

I assert that political polling is a critical component of any democratic society. By measuring public opinion on a range of political issues and candidates, surveys, and polls provide valuable insights into the minds and motivations of the citizens. From tracking voting patterns to predicting election outcomes, political polling has become a crucial tool for understanding what drives voters and for shaping the political landscape.

At its core, political polling is about understanding people and their motivations. By using scientific sampling methods to get a representative slice of the population, pollsters can analyze the data to see what issues and policies resonate with different groups of voters. Whether it's tracking changing opinions on immigration, gun control, or the People's Republic of China, political polling offers a way to gauge how public opinion shifts over time.

However, the impact of political polling extends far beyond election cycles and policy agendas. Poll results can also *influence* public discourse and shape the way news outlets report on political issues. When a poll shows that a particular candidate is ahead in the polls, for example, it can lead to increased media coverage and fund-raising support, which, in turn, can impact the dynamics of the race.

Moreover, political polling is integral to the democratic process itself. By offering a window into what people are thinking and feeling, polls provide valuable feedback to elected representatives and policymakers. As political parties (and interest groups) strive to find effective messaging and outreach strategies, polling data can help them identify which issues are most important to voters and which arguments are most persuasive.

Of course, political polling is not without its limitations and challenges. From issues with sample biases to low response rates, pollsters must navigate a complex array of factors to produce reliable—and accurate—results. Additionally, the rise of social media and other digital platforms has created new opportunities for influence, making it increasingly difficult to separate actual public opinion from propaganda or targeted marketing campaigns.

Despite these limitations and challenges, political polling remains a critical tool for understanding the dynamics of political power and the complex interplay of public opinion and policy. Whether it is breaking down demographic shifts or tracking changing attitudes toward climate change or racial justice, polling data provides a unique and valuable perspective on the issues that shape society.

Ultimately, political polling is about more than just numbers on a page; it is about understanding the complex intersections of power, identity, and ideology that shape the political world around us. By exploring the implications of political polling in the digital age, we can gain a deeper appreciation for the ways in which data can be used to empower or marginalize communities, reinforce or challenge the status quo, and pave the way for a more equitable and just society.

This book takes us on a journey covering the history, the practice, and the impact of political polling on the modern political landscape. By delving into case studies and in-depth analysis of polling techniques and methodologies, we will shed light on the inner workings of this essential tool and the role it plays in shaping the world around us.

1.2 Polling Terms

But, before we can complete this journey, we need to take the first steps. As with any new discipline, those first steps tend to focus on terminology.

1.2.1 The Populations and the Sample

The core of all applied statistics is the **sample**. The sample is the data that was measured. The sample is used to better understand the target population (a.k.a. the **population of interest**). By definition, the sample is a subset of the **sampled population**, which may or may not be a subset of the target population. Figure 1.2 shows a schematic of the relationships among the target population, the sampled population, and the sample. To be clear, the sampled population is not necessarily a subset of the target population. This is especially true in political polling.

In the realm of polling, the population of interest is those who *actually* cast a valid vote in the election. The sample is those who are successfully surveyed by the polling firm. The sampled population is the group from which the sample is drawn. Since polling firms tend to want their sampled population to be as close to the target population, they tend to limit their sample to "likely voters"—at least later in the election cycle. Early in the cycle, it is rather difficult to know who will and who will not vote, so they tend to include registered voters (or even all adults).

The Purpose of the Sample

There is one—and only one—purpose for the sample. It is to represent the population. A sample that fundamentally differs from the population will lead to improper conclusions (but see Chapter 4).

Unfortunately, one can never know the voting population before the vote; in fact, it does not even *exist* before the vote. Thus, this is one source of error in predicting election outcomes.

The box above mentions that not knowing the population leads to estimation errors. This is because the polling house must make assumptions about the turnout. If the composition of the voting population matches what the polling firm supposes, then the predictions will be rather close.

However, if one segment of the population turns out at a surprising rate, then the estimates will be wrong. In election studies, when a segment of the

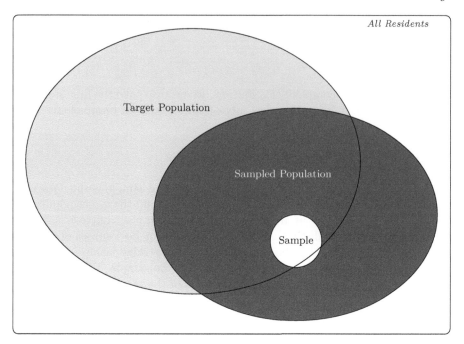

FIGURE 1.2
Illustration of the relationship among the target population, the sampled population, and the sample. Note that the sample *must be* a subset of the sampled population, but that the sampled population does *not* have to be a subset of the target population.

population turns out at a higher rate than expected; we say that the election "broke" for that candidate. We then look back over our data to see if there was evidence that this would happen. Usually, there is not.

1.3 Sampling Methods

The sole purpose of the sample is to represent the population in "important" ways. Usually, the sample should match the population in terms of political beliefs. Thus, if the voter support for Lai of the DPP is 35%, then it would be important for the sample to estimate support rate close to this. If 45% of the females support Lai, then the sample should have about 45% of its females support Lai.

The problem is that we do not know the actual level of voter support (or of female voter support) for a candidate until the election actually occurs. As such, we can never actually know if our sample is representative of the population in "important" ways. If we knew this in the population, then there would be no need to do statistics, and I would be out of a job.

To help ensure that our samples *are* similar to the target population, statisticians have created several legitimate sampling methods that can be used. Unfortunately, each has strengths and weaknesses. The next sections briefly explore some of the main sampling schemes.

1.3.1 Simple Random Sampling (SRS)

In simple random sampling (SRS), every member of the sampled population has an equal probability of being selected for the sample. The mathematics of SRS are the easiest to understand, so we will spend an entire chapter examining this sampling method (Chapter 2).

Its simplicity is its strength. Statistically, its simplicity also leads to these estimators being unbiased. Being unbiased is actually not that impressive. It does not mean that the estimated value is close to the population value. It *does* mean that if we sample a gazillion times, then the average of the calculated estimates will be the population value. Thus, unbiasedness is a large-scale property; it says nothing about a single estimate.

The main weakness of simple random sampling is that its estimates are unstable. In other words, if you and I collect different samples and estimate candidate support on each, our estimates may be *very* different. High instability reduces the legitimacy of a single estimate.

Later in this chapter (page 14), we will use R to explore this instability in our first statistical experiment.

The Purpose of Experiments

In the sciences, one experiments under various conditions to better understand the relationships between and among the variables. The same should be done in statistics. Through experimentation, one better learns about the effects of randomness on estimates.

One quality of this book that I like is that I provide several experiments throughout so that you can gain some additional insight into the statistics of polling.

1.3.2 Stratified Sampling

When polls are reported, readers will typically *think* in terms of SRS. The experienced ones will look at the **cross-tabs** to examine the make-up of the sample (Section 4.3). They will then critique the poll for having, for instance, 75% female, 50% DPP, and 15% age of 18–29. Since it is highly unlikely that the proportion of females in the population is 75%, they will claim that this poll is skewed toward the "female position."

If the polling firm *is* using SRS, then this criticism is *entirely* appropriate. Those sample proportions are not even close to the expected population proportions. However, the vast majority of polling firms use a sampling scheme called "stratified sampling." The advantage of stratified sampling is that the estimates are not as unstable as the SRS estimates. It has disadvantages, however.

Stratified sampling is a type of weighted sampling. It uses **weights** (multipliers) to adjust the sample so that it more closely matches the population of interest. If the population has 50% female and the sample has 75% female, then female responses are down-weighted by a third to ensure they count 50% of the final estimates. The math is straightforward:

$$75\% \times \left(1 - \frac{1}{3}\right) = 50\%$$

The math is straightforward. Knowing the correct proportion is not. How does one determine the actual proportion of voters who are female? Should the proportion be that of the registered voters? Should it be "likely" voters? Should it be the proportion from the previous election?

Chapter 4 explores stratified sampling in greater detail. This detail includes the mathematics. More importantly, it includes a discussion (and experiment) to see how close you have to be to be better than simple random sampling.

1.3.3 Non-Probabilistic Methods

The previous methods rely on planned randomness. Simple random sampling requires that every member of the sampled population has an equal chance of being selected. Stratified sampling requires every member of each stratum to have an equal chance of being selected.

However, there are methods that rely on the judgment of the researcher. The researcher will seek out participants based on their understanding of the population of interest. For instance, if I knew a certain segment of the population would tend to be missed by the above probabilistic sampling methods, I may purposely seek out members from that group and include them in my sample to help ensure it is more representative of the population.

The drawback to using such **purposive sampling** methods is that I may *not* actually understand the target population. Thus, in seeking out members of that group and including them in the sample, the sample may lose its one goal: to be representative of the population of interest.

Planned Samples and Panels

Since the turn of the millennium, there has been a growing crisis in polling. Larger and larger segments of the population are not being represented in the samples. To fix this issue, some polling firms are jettisoning probabilistic sampling in favor of planned samples.

For instance, the University of Southern California's Dornsife Center for Economic and Social Research selected a sample of approximately 3000 people in the run-up to the 2016 US presidential election. They would poll this curated sample weekly and weight the results to match the population of the United States.[Dor16] This particular example of non-probabilistic panel sampling turned out to be no better or worse than most other samples. However, what we learned from the experience was quite valuable for polling research. They used their research to improve their work for the 2020 election.[Dor20] They missed 2016 by 5 percentage points (pp), the 2020 by 2pp. Thus, there was a marked improvement in their forecasting. Was this due to their changes or to a change in the electorate?

Convenience Sampling

Beware of convenience sampling and non-probabilistic schemes designed to push a point. Actually, in general, be wary of any poll that is trying to drive the narrative. The purpose of legitimate polls is to *understand* the population, not to affect it. This is why I refer to this as an "evil sampling method" in my classes.

1.3.4 Other Sampling Methods

Of course, there are other sampling schemes. There are also variations on the above schemes to take advantage of well-known population features (if such exists). However, I think these are beyond the scope of this book. Understanding the above three methods will be enough for 99.9% of polls. The fancier methods will tend to be seen in academic literature much before being used "in the wild."

1.4 Biases in Polling: The First Words

Bias in polling refers to systematic errors or distortions in survey results that result from factors other than random chance. These biases can arise due to various sources, such as sampling methods, question wording, response rates, and non-response bias. There are several sources of bias in a study, broadly divided into two:

- researcher-based
- respondent-based

By the way, **no study is perfect**. No sample will perfectly represent the population. Bias is only important when it is sufficiently severe to cause the polling analyst to draw incorrect conclusions.

1.4.1 Researcher-Caused Bias

One group of biases occurs because of the researcher. At no point should we consider the researcher is causing this bias *deliberatively*. Such an action would be unethical (Section 1.5). However, these biases creep into a study if one is not careful. Having a better understanding of these sources allows you to think about how the data were collected and whether that has an impact on the conclusions drawn.

Selection Bias

This occurs when the *selection* of individuals for the poll is done in such a way that proper randomization is not achieved. This is pervasive in polling, especially in single-mode polling. For instance, if your poll consists of contacting people with a landline telephone, then you will exclude a large portion of the population from your poll. If, additionally, those with landlines vote differently than those with smartphones, you have a problem.

Response Bias

This occurs when a researcher's behavior (or mere existence) causes a participant to alter his or her response. This could be caused by something as minor as the researcher's tone of voice, clothing choices, race or ethnicity, or demeanor.

Popular psychology tells us that humans want to belong. Most of us want to make the other person happy—at least while we have to deal with them. Thus, we will subconsciously alter our responses to match what we think their expectations are. Because of this, surveyors dressed in three-piece suits may tend to have a higher proportion of conservative responses, while those wearing "flower power" clothes will have a higher proportion of "liberal" responses in their sample.

Note that this may not be a large factor, and its effect will (most likely) be drowned out by larger sources of bias. But, it is important to understand why call centers (the ones who actually telephone the person to survey) tend to use neutral tones and follow a well-rehearsed script.

1.4.2 Participant-Caused Bias

Beyond conscious biases, there are several that do not arise from planning errors. They arise from the actions of the participants. The most important in terms of polling is participation bias.

Participation Bias

This occurs when there is a problem with the participation—or lack thereof—of those chosen for the study.

As an example, **nonresponse bias** occurs when there is a lack of participation in a self-selected sample from certain segments of a population or when a person refuses to participate in a survey. This is especially pernicious if the non-responders tend to hold a political position different from the responders. For instance, if 80% of those who support Candidate A decide to never answer polls, while the supporters of Candidate B always answer polls, then polls will be biased toward Candidate B.

Dropouts

These are participants who begin a study but fail to complete it. This term is usually applied as a type of non-response bias, but it is important enough to have its own section.

Dropouts result in surveys that have **missing data**. There are a few was of addressing missing data, but they depend on knowing that the data are missing at random, that is, that the missing values are not related to what is being measured.

If the missing data *is* related to candidate support, then there is a fundamental problem with the survey. If the missing data is important, then the best—but most expensive—option is to redesign the entire survey and start anew.

1.4.3 More Bias in Sampling

In the concluding chapter of this book, Chapter 13, I return to the question of bias because it is so important to understand. However, if you are interested in a better understanding of biases in research, I encourage you to study Table 1 in the Berkman et al. report. In that table, they provide a nice list of possible sources of bias and how to solve them. While the list is for medical research, many are applicable to polling research.[BSVM14]

1.5 Legitimacy in Polling

In any election cycle, there are many polls conducted. Usually, a media company will pay a polling firm to do the poll and the analysis. Frequently, *that* polling form will pay a company to collect the data for the analysis. While legitimacy must exist at all three levels for the poll to be legitimate, I believe that the emphasis should be on the legitimacy of the *polling firm*. They are responsible for the proper collection, analysis, and presentation of the data.

So, how do we know if a polling firm is legitimate?

The American Association for Public Opinion Research (AAPOR) is the primary association of survey researchers in the United States.[AAP24c]

> The AAPOR community includes producers and users of survey data from a variety of disciplines. Our members span a range of interests including election polling, market research, statistics, research methodology, health-related data collection and education.

They have developed a code of ethics and practices that all polling firms *should* follow, whether they are a member of AAPOR or not. This code outlines the ethical responsibilities of all public opinion researchers. I encourage you to read the "**AAPOR Code of Professional Ethics and Practices**." Doing so will give some insight into the amount of thought we give to correctly present public opinion to our audience.[AAP24a]

https://aapor.org/standards-and-ethics/

In addition to the code, AAPOR has spearheaded an initiative designed to promote transparency in public opinion polling.[AAP24b] According to them,

> AAPOR's Transparency Initiative is designed to promote methodological disclosure through a proactive, educational approach that assistssurvey

organizations in developing simple and efficient means for routinely disclosing the research methods associated with their publicly released studies.

The Transparency Initiative is an approach to the goal of an open science of survey research by acknowledging those organizations that pledge to practice transparency in their reporting of survey-based research findings. In doing so, AAPOR makes no judgment about the approach, quality or rigor of the methods being disclosed.

Thus, I would assert that a sufficient requirement for legitimacy of a polling firm is to be a member of this initiative. Since there is not cost to join, I think all polling firms should join and strengthen the quality of public opinion research in the United States.

Since this began in 2014, the number of member institutions has grown. Current members include ABC News, East Carolina University's Center for Survey Research, and Marquette University Law School Poll. In my own research, I will not use a poll from a firm that is not a member of the Transparency Initiative.

Beyond AAPOR's Borders

Even though the American Association for Public Opinion Research has "American" in its name, membership is not limited to American firms. Regardless, there are other such professional organizations in the world. Some of the larger ones are

- European Society for Opinion and Marketing Research (ESOMAR)

  https://esomar.org/

- European Survey Research Association (ESRA)

  https://www.europeansurveyresearch.org/

- World Association for Public Opinion Research (WAPOR)

  https://wapor.org/

I encourage you to investigate the ones for your area. They all have codes of ethics that their members are supposed to follow. This shows that true survey research is a professional activity.

1.6 Experiments in Polling

In my humble opinion, statistics is a science and not a mathematics. Because of this, I encourage my students to experiment. Since I now claim you as one of my students, I encourage *you* to experiment. Experimentation in statistics requires a computer and software to run the simulations. My choice of software is the R Statistical Environment.[R 23] If you are not familiar with R, please visit Appendix A. In that appendix, I walk you though the installation of Rand provide a few examples of statistical analyses using it.

Now that you are familiar with R, let us do a statistical experiment. In this experiment, we will generate random samples from a known population and estimate the population proportion for each. The purpose of doing this is two-fold. First, it is an easy experiment to do in R. Thus, it is a good first experiment. Second, it shows the variability of simple random sampling.

Let us begin.

Experiment: Let us start with a population. In this population, 53% support Candidate F for president. Now, let us take a random sample of size $n = 500$ from this population:

```
x = rbinom(500, size=1, prob=0.53)
```

Note that each element of that sample will either support Candidate F for president or not. The former is identified with a 1; the latter, 0. To see the results of our poll, run this

```
x
```

When I ran this, R printed out the following:

```
  [1] 1 1 1 0 0 0 0 1 0 1 1 0 1 1 1 0 1 1 1 1 0 0 1 1
 [25] 0 0 1 1 0 0 1 0 0 1 0 0 1 1 0 0 0 0 1 1 0 1 0 1
 [49] 0 1 0 1 0 0 1 0 1 1 1 0 0 1 0 0 0 0 0 1 1 1 1 0
 [73] 1 0 1 1 0 0 1 1 0 1 1 1 1 1 1 1 1 1 0 1 1 0 1 0
 [97] 1 1 1 1 1 1 1 0 0 1 0 0 0 1 0 0 1 1 1 0 1 0 0 1
[121] 1 1 1 0 0 0 1 0 1 1 1 0 1 1 0 1 0 1 1 0 0 0 0 0
[145] 0 1 1 1 0 1 0 0 0 0 0 1 0 1 0 0 1 1 1 0 1 0 0 0
[169] 1 0 1 1 1 0 0 1 1 1 0 1 0 0 0 0 1 1 1 1 0 0 0 0
[193] 1 0 1 0 1 1 0 1 0 1 1 0 0 0 0 1 0 1 1 1 1 1 1 0
[217] 1 1 0 0 1 0 0 0 0 0 0 1 0 0 0 1 0 0 0 0 1 1 1 1
[241] 1 0 0 1 0 1 1 1 0 0 0 1 1 0 0 0 1 1 1 1 0 0 1 0
[265] 1 0 1 1 0 1 0 1 1 0 1 0 1 0 0 1 1 0 0 1 1 1 0 1
[289] 1 1 1 0 0 0 0 1 0 1 1 0 0 1 1 0 0 0 1 0 0 0 1 1
```

Experiments in Polling

```
[313] 0 1 0 0 0 0 0 1 1 0 0 0 1 1 1 1 1 1 0 1 1 1 1 1
[337] 0 1 0 0 0 1 1 1 0 0 0 0 1 1 1 1 1 0 1 1 0 1 1 0
[361] 1 0 1 0 0 1 0 0 0 1 0 0 0 1 1 0 0 0 1 0 1 0 0 0
[385] 1 1 1 0 1 1 0 0 0 1 0 1 0 1 0 1 1 1 1 1 1 1 1 1
[409] 1 1 1 1 0 0 0 0 1 1 1 1 0 0 0 0 0 1 0 1 1 0 0 0
[433] 0 0 0 1 1 1 1 0 1 0 1 1 1 1 1 1 1 1 0 1 0 0 0 1
[457] 1 0 0 0 0 1 1 1 0 1 1 1 0 0 1 1 1 0 1 1 0 0 0 0
[481] 1 0 1 1 0 1 0 1 0 1 1 0 1 1 1 1 0 1 0 1
```

Each 0 and 1 represents a response to the poll. There are 500 of them here because we specified that there would be 500 people sampled.[2] In this sample, 264 stated that they supported Candidate F. This means that the sample proportion is $264/500 = 0.528 = 52.8\%$.

Note that this is *not* equal to the known support level in the population, $\pi = 0.53$. This should not surprise us because the sample is random, so the results will also be random.

Since we care only about the number of people in the sample who support Candidate F, and not *which* ones, we can also run the following line

```
rbinom(1, size=500, prob=0.53)
```

Here, the 1 represents the number of polls done, the 500 indicates the sample size of the poll, and the 0.53 is the support for Candidate F *in the population*.

Running this line, I got 256. This differs from the 264 I got in the previous code *and* it probably differs from what you got. Why? The sample is random, so the results will also be random.

Since there were 256 in my sample who supported Candidate F, the estimated support for Candidate F *in the population* is $256/500 = 51.2\%$. This is different from my first estimate and from the real value. It is also different from your two estimates. Why? The sample is random, so the results will also be random.

Did You Know?

If you take only one thing away from this experiment, I hope it is this:

The sample is random, so the results will also be random.

This statement explains why two polling forms can perform the same poll and come up with different results. It explains much of the variation in

---

[2] By the way, your output will (most likely) differ from mine. When polling, the sample is random. Thus, the series of 0s and 1s will be random.

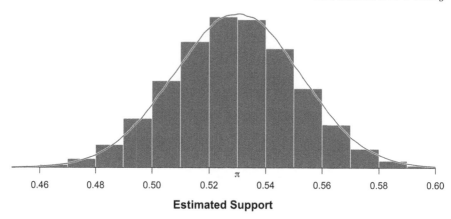

FIGURE 1.3
Histogram (with overlaid density curve) for a million synthetic polls estimating the population support for Candidate F. The actual population support is represented by the π along the axis.

reported polling results. It is also a source of frustration for researchers and readers alike.

And now for the beauty of this experiment. Remember that we would like to illustrate the instability of simple random sample estimates of population support. Above, we generated two synthetic (fake) polls and estimated the population support for Candidate F. There was clearly some variability in the estimates (they were not the same). To get a better understanding of this variability, we will need generate the results from a great many synthetic polls.

The following code performs simple random sampling a million times (1×10^6, 1e6) and calculates the proportion (p) for each:

```
x = rbinom(1e6, size=500, prob=0.53)
p = x/500
```

Now, we have the estimated proportions from a million synthetic polls.

Let us first *look at* the variability. Figure 1.3 is a histogram (with overlaid density plot). The actual population support is represented by π along the axis. It is at 53%.

While the estimated supports do cluster around $\pi = 53\%$, there are quite a few relatively far away from the right answer. A significant proportion of

Conclusion and Looking Ahead 17

the polls actually have Candidate F's support below 50%, which would be problematic in a real election. (Why?)

The following code calculates some useful statistics

```
mean(p)
mean(p<0.50)

quantile(p, c(0.025,0.975))
sd(p)
```

These statistics tell us the following: First, the average (mean) of all of those polls is actually 53%. Thus, this sampling method is "unbiased" (see Section 2.2.1). Even so, 8.3% of those polls had Candidate F losing the election. To illustrate the variability in the poll estimates, 95% of the polls show support between 48.6% and 57.4%. By the usual definition, "reasonable" polls will fall in that interval.

The standard deviation of the polls estimates is 2.23 percentage points. In itself, this number is not too useful, especially since we already calculated the bounds on the middle 95% of the estimates. However, the standard deviation can be used to compare two estimators to determine which is the more precise of the two.

◇ ◇ ◇

Section 2.2.1 explores bias of estimators. There, we see that bias is a good thing to have in an estimator. However, further reading suggests that it is not the most important thing. Section 2.2.2 looks at estimating variability (instability). Lower instability indicates that the estimates are more precise, that they vary less from sample to sample. Most importantly, Section 2.2.3 explores one method for combining the bias and variance into a single measure of estimator concentration about the value being estimated, π. It is this measure that we want smallest in our estimators.

1.7 Conclusion and Looking Ahead

If we were to reduce this chapter to two "bumper sticker" sayings, the first is

- **the target population does not entirely exist until election day.**

This means that the sampled population may or may not be a subset of the target population. It also means that stratified sampling cannot be done perfectly.

Second, and more importantly from the standpoint of understanding:

- **The sample is random, so the results will also be random.**

This explains why two good, legitimate polls will result in different estimates. While the variability is understood and can be modeled, it cannot be eliminated.

And so, this chapter serves as a first step down the long road of understanding elections through statistics. This first step introduced some terms that will be seen and used frequently in the pages ahead.

1.8 Extensions

The chapter is almost over. However, I cannot let the opportunity pass to give an assignment. Since this chapter really just introduced some basic terms, there is not much to do. But, here are some questions.

Review

1. Why is political polling important?
2. Why are no political polls exactly correct in their predictions?
3. What are the differences between the target population and the sampled population? What would these be in a political poll?
4. What is the purpose of the sample?
5. Which of the sampling methods covered is unbiased?

Conceptual Extensions

1. Describe the geographical separation in Taiwan between the DPP and the KMT.
2. In political polling, the target population is "People who cast a ballot in the election." Explain why it is impossible to ensure that the sampled population is a subset of the target population... until after the election.

Computational Extensions

1. Determine if the variance of the estimator used in this chapter depends on the candidate's support in the population. To do this, compare the standard deviations for the following values of the population support: $\pi = 0.50, 0.25, 0.10, 0.01$.

1.9 Chapter Appendix

This appendix contains some "review" material for this chapter. For this chapter, it just provides the R commands used and explains them briefly. In other chapters, the chapter appendix will contain proofs, additional experiments, and the like.

1.9.1 R Functions Used

A statistical program is invaluable in gaining a better understanding of statistics and its/their relationships to the world around us. The following R functions were featured in this chapter.

- `hist`

  This produces a basic histogram of the provided numeric variable. As with all R graphics functions, there are enough options to create just about *any* histogram your heart desires.

- `mean`, `sd`

  These calculate the arithmetic mean and the standard deviation of the values in the x variable.

- `quantile`

  This calculates the specified quantile(s) of a vector of numeric values. For instance, `quantile( pollResults, c(0.025,0.975) )` calculates the 2.5 and 97.5th percentiles of the values in the `pollResults` variable.

- `rbinom`

  All basic distributions are represented in R. For each, there are usually four things that can be calculated on the distribution: likelihood, cumulative probability, quantile, and a random value. The structure of R probability functions is a prefix, denoting what needs to be calculated, and a stem, denoting which distribution.

Here, the `binom` indicates a Binomial distribution is being used, and the `r` indicates random values are being generated. That means that this line `rbinom(10, size=500, prob=0.53)` generates 10 random values drawn from a BINOM($n = 500, \pi = 0.53$) distribution. In the text, this distribution represents the outcome of a single poll in which 500 are asked and the *population* support is 53%.

2

Simple Random Sampling

Firms produce a myriad of polls each election cycle, with each poll attempting to accurately estimate the proportion of the voting population holding a given position. For instance, during the lead-up to the 2014 referendum on Scottish independence, there were well over 100 polls taken. The polls were not consistent; their estimated support for independence ranged from 25% to 49%. In fact, even those polls taken in final week of the campaign varied in their estimated support from 41% to 47%.

In this chapter, we will begin our study of elections by focusing on obtaining information from opinion polls. Here, we will take the simplest polling scheme and build the statistical foundation for a deeper understanding of polls, how they are performed, and what they can tell us. Along the way, we will come to a deeper understanding of the causes of that variation, as well as its effects.

Scotland: The Independence Referendum of 2014
While the theory of the nation-state arguably dates from the early nineteenth century, the idea of a separate Scotland dates from at least the Middle Ages, echoing its separation from (and frequent battles with) England.[Dev08] It was the personal union between the two kingdoms (1603–1707) that set the stage for an uneasy legal union between the two: The Treaty of Union of 1707, which joined the Kingdom of Scotland to the Kingdom of England to form Great Britain.[Dev08] Since then, the union has gone through phases of strength and weakness, of unity and conflict. With the rise of the ideal of the nation-state, Scots began agitating more strongly for independence. From the mid-nineteenth century onward, the United Kingdom began a series of moves to place more local power in the hands of the Scots to keep the United Kingdom whole.

That we are still discussing Scottish independence as being a (possible) future event strongly points to the success of devolution. However, Scottish independence remains an issue in the United Kingdom, with the Scottish National Party (SNP) being a significant power in the Scottish parliament. With the success of SNP and its pro-independence platform, it was inevitable that Scotland would vote on its independence from the United Kingdom.

In preparation for the independence vote, the Scottish parliament (a.k.a. Holyrood) passed the Scottish Independence Referendum Act (2013 asp 14).[Sco13] This act set the terms of the independence referendum, which included who would be allowed to vote, when they would vote, and what they would vote upon:

"Should Scotland be an independent country?"

On September 18, 2014, a total of 1,617,989 votes were cast in favor of independence—44.70% of the total counted.[MAH14] It is interesting to note that neither the support for independence nor the turnout was uniform throughout Scotland (see Figure 2.1). It is also interesting to note that there was no clear urban-rural divide in independence support. For instance, Glasgow supported it at 53.5%, while equally urban Edinburgh supported it only at 38.9%.

The results were not entirely surprising. Over 100 opinion polls were taken regarding Scottish independence, with only a few showing independence being in the lead. One of those polls was taken in the waning hours of the referendum campaign. On September 16 and 17, Survation polled 1266 Scots, generating 1061 effective responses.[Sur14] They contacted landline and mobile phones. They weighted the results according to age, sex, region, previous party vote for the parliament, and voting likelihood. Their poll estimated the support for independence to be 47%.

Why was this sophisticated-sounding poll wrong with its estimate? Or, *was* it wrong? Is being off by 2.3% "close enough" for a poll?

Overview

Ultimately, the purpose of any legitimate poll is to estimate the proportion of the target population in favor of some position.[1] However, since only a small group (the sample) is polled, one can never be absolutely certain about the position of the population. Thankfully, statistics provides several tools

---

[1] There are illegitimate polls whose purposes are to influence a person's position. These are ignored in this monograph as being entirely unethical. The fundamental purpose of an ethical poll is to understand the population, not *influence* the population.

Overview

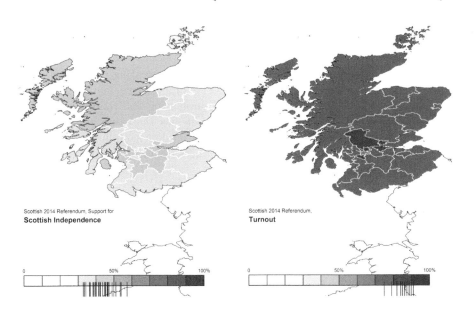

FIGURE 2.1
Result maps of the 2014 independence referendum for Scotland. The left map illustrates the proportion of votes in favor of independence by local authority. The right map illustrates the turnout, also by local authority. These data are from Wikipedia.[Wi20b]

allowing us to draw some conclusions about the population based on that sample.

One of the simplest methods for properly conducting a poll is to contact a large number of people and ask their position on a question. For instance, Survation could have called 1061 residents of Scotland and asked them how they planned to vote (in favor of, opposed to, or uncertain about independence). They would then use the proportion of their sample supporting independence as their estimate of the population supporting independence.

In reality, there is a bit more to it than that. However, to allow us to gain a feel for what the polls actually tell us, this chapter simplifies polling. In Chapter 4, we increase the complexity of the polling scheme to better match reality.

2.1 Simple Random Sampling (SRS)

A **sampling scheme** is a structured method for generating (or collecting) a sample from the **sampled population**. The simplest scheme for polling is to have a list of the people in the sampled population and randomly contact people on that list, ensuring that each person has the same probability of being contacted. This method is called "simple random sampling," (SRS), the list of people in the population is termed the **sampling frame**, and those contacted compose the **sample**.

In SRS, each person in the sampled population has the same likelihood of being contacted by the polling firm. In fact, at each contact point, each member of the sampling frame has the same probability of being contacted. While this leads to the possibility of a person being contacted multiple times, it also makes the problem much more mathematically tractable, as you will see.

To wit: Let us define π as the proportion of the target population holding a specified voting preference, whether a candidate or a referendum position. Let n be the number of people surveyed (the sample size). Finally, let X be the number of people *in a particular sample* who hold the given position. Then, under the simple random sampling scheme, X is a random variable that follows a Binomial distribution. We write

$$X \sim \text{BINOM}(n, \pi) \tag{2.1}$$

Completely specifying the distribution of the data is extremely important in statistics. Among other things, it allows one to fully understand a cause (a source) of the variation in the observed values. Since the fundamental goal of polling is to use this information to properly estimate π, knowing the distribution of the data allows us an opportunity to estimate that parameter. This is the goal of the next few sections.

2.2 One Estimator of π: The Sample Proportion

While the parameter of interest, π, is unknown, one can easily use probability statement (2.1) and a knowledge of the Binomial distribution to create an estimator of π (Appendix B.2). Because the expected value of a Binomial random variable is $\mathbb{E}[X] = n\pi$, the sample proportion is an obvious choice,

$$P = \frac{X}{n}$$

One Estimator of π: The Sample Proportion

There are at least two benefits to using the sample proportion as the estimator.[2] The first benefit is that it just makes sense to use: "The sample proportion estimates the *population* proportion." The second benefit is that the sample proportion is an **unbiased estimator** of π (see Section 2.2.1).

Quick Example

In the Survation survey above, 496 of the 1061 polled stated that they were voting in favor of independence. Thus, the sample proportion estimate for the support for independence is $p = x/n = 496/1061 = 46.75\%$.

Note that the sample proportion is *an* estimator of π. There are other estimators that can be used. Because there are many estimators, it will be helpful at times to compare estimators, determining which is/are superior to others—and when. There are three usual ways to compare estimators: bias, variance and, mean squared error. The next three sections examine these measures.

2.2.1 Estimator Bias

Above, I wrote that $P = X/n$ was an "unbiased estimator of π." What does this actually mean? To be an unbiased estimator, the expected value of that estimator must equal the parameter of interest. The following definition is more mathematical, but says the same thing.

Definition 2.1. *Let P be an estimator of π. The **bias** of P is defined as*

$$\text{bias}[\,P\,] = \mathbb{E}[\,P\,] - \pi$$

*We say P is **unbiased** for π if $\text{bias}[\,P\,] = 0$, i.e., if $\mathbb{E}[\,P\,] = \pi$.*

With that definition, it is straightforward to prove that the sample proportion is an unbiased estimator of π (frequently spoken as "P is unbiased for π"). To do this, we need only calculate the expected value of P and compare it to the parameter π:

Theorem 2.2. *Define $P = X/n$, where $X \sim \text{BINOM}(n, \pi)$. P is an unbiased estimator of the population proportion, π.*

---

[2] For those interested, the sample proportion estimator is *both* the Method of Moments estimator and the Maximum Likelihood Estimator for the population proportion, π. While there are many other methods for creating estimators, Method of Moments estimators tend to produce estimators that seem reasonable (see, e.g., [Ric07]).

Proof. Given $P = X/n$, where $X \sim \text{BINOM}(n, \pi)$, we have the following:

$$\mathbb{E}[\,P\,] = \mathbb{E}\left[\frac{X}{n}\right]$$

$$= \frac{\mathbb{E}[\,X\,]}{n}$$

$$= \frac{n\pi}{n}$$

Thus, $\mathbb{E}[\,P\,] = \pi$

Since $\mathbb{E}[\,P\,] = \pi$, Defn. 2.1 tells us that P is an unbiased estimator of π. In other words, $\text{bias}[\,P\,] = 0$.

∎

The advantages to an unbiased estimator are clear. Are there disadvantages? Frustratingly, the usual statistics answer is "it depends," as the next few sections suggest.

Another Estimator: Agresti-Coull

The sample proportion is not the only estimator of the population proportion, π. A different estimator was created by Professors Agresti and Coull.[AC98] The eponymous Agresti-Coull estimator is defined as:

$$\hat{\pi}_{ac} = \frac{X+1}{n+2}$$

While one may think that Agresti and Coull pulled this estimator from thin air, it actually arises from two separate sources. The first is from Laplace's Law of Succession.[Pie14, Wil27] The second is from using Bayesian analysis and a standard Uniform prior distribution (Chapter 5).

I will let you read the Laplace work; it is well-worth the read if you enjoy the historical aspects of statistics and how we tried to answer these questions in ages past. The original Agresti-Coull paper is also quite interesting, too.[AC98] These two works are separated by almost two centuries, so they offer an explicit view at how statistics has changed.

The point of discussing the Agresti-Coull estimator here is to give some more insight into the "mathematics of statistics" and to offer another option to see how the bias is calculated. By the end, you will be able to conclude that he Agresti-Coull estimator is biased (unless $\pi = 0.500$).

$$\text{bias}(\hat{\pi}_{ac}) = \mathbb{E}[\,\hat{\pi}_{ac}\,] - \pi$$

One Estimator of π: The Sample Proportion

$$= \mathbb{E}\left[\frac{X+1}{n+2}\right] - \pi$$

$$= \frac{\mathbb{E}[X]+1}{n+2} - \frac{\pi(n+2)}{n+2}$$

$$= \frac{n\pi+1}{n+2} - \frac{n\pi+2\pi}{n+2}$$

Simplification gives us

$$\operatorname{bias}(\hat{\pi}_{ac}) = \frac{1-2\pi}{n+2}$$

Note two things about this estimator.

- First, as the sample size increases, the bias converges to zero. That is, the Agresti-Coull estimator is asymptotically unbiased.
- Second, if $\pi = 0.500$, the estimator is unbiased. To see this, substitute $\pi = 0.500$ into the bias result and simplify.

At this point, one must wonder why Professors Agresti and Coull spent time creating a biased estimator *and* why I am spending time talking about them. Bias is not the entire story, as the next section discusses.

What Bias Tells Us

What we *really* want from our estimators is to have them produce estimates that are always close to the real value. Unfortunately, bias tells us absolutely *nothing* about whether a single estimate is close to the true value. It only tells us about its "large-scale" averages.[3]

An unbiased estimator is only guaranteed to give the right result if the experiment is repeated an infinite number of times and the individual results are averaged. That is, "bias" speaks to the performance of the estimator as a whole. It tells us precious little about the individual estimates.

To see this, imagine you use some estimator of the population proportion and obtain a value of 0.574. Even if the estimator is unbiased, there is no way to know if that particular estimate is close to reality or far from it. If the estimator is highly **concentrated** about the true value, however, we then know that the estimate of 0.574 is close to reality.

A first attempt at understanding concentration is the variability (a.k.a. variance, instability) of the estimator. We cover that next.

---

[3] This is true because bias is defined based on the expected value, which is just a long-term average of outcomes.

2.2.2 Estimator Variability

The above suggests that **consistency** (a.k.a. precision) is also important in an estimator.[4] If we repeat an experiment a gazillion times, then it would be nice if the estimates are close to each other. This would indicate that the estimator is precise, that the uncertainty created by using this particular estimator was low. So, all things being equal, we prefer an estimator with a lower variance.

Let us calculate the variances of the two estimators. In variance calculations, remember that the variance of a single number is 0 and that $\mathbb{V}[\,aX\,] = a^2\,\mathbb{V}[\,X\,]$ (factoring out a constant requires the constant to be squared).

First, the variance of the sample proportion.

$$\mathbb{V}[\,P\,] = \mathbb{V}\left[\frac{X}{n}\right]$$

$$= \frac{\mathbb{V}[\,X\,]}{n^2}$$

$$= \frac{n\pi(1-\pi)}{n^2}$$

$$\text{Thus, } \mathbb{V}[\,P\,] = \frac{\pi(1-\pi)}{n}$$

Note three things from this result:

- The precision of P increases (variance decreases) as n gets larger. In other words, more data gives us more information about π.

- The estimator is least uncertain (most precise) when $\pi = 0$ or $\pi = 1$. If we think of X as being the outcome of flipping a coin, then this means we are most sure about the outcome when the coin is two-tailed ($\pi = 0$) or two-headed ($\pi = 1$).

- The estimator is *most* uncertain (least precise) when $\pi = 0.500$. This means a fair coin produces the greatest uncertainty in the outcome of the flip.

The Variance of the Agresti-Coull

Next, let us calculate the variance of the Agresti-Coull estimator so we can compare it to that of the sample proportion.

$$\mathbb{V}[\,\hat{\pi}_{ac}\,] = \mathbb{V}\left[\frac{X+1}{n+2}\right]$$

---

[4]In theoretical statistics, the term "consistency" means something specific. Here, I am using it in the colloquial sense.

$$= \frac{\mathbb{V}[\,X+1\,]}{(n+2)^2}$$

$$= \frac{\mathbb{V}[\,X\,]}{(n+2)^2}$$

$$= \frac{n\pi(1-\pi)}{(n+2)^2}$$

The three things we noted about the sample proportion estimator are the same for the Agresti-Coull estimator. The interesting part is that the variance of the Agresti-Coull estimator tends to be *lower* than that of the sample proportion. To see this, let's compare the two and see when they are equal:

$$\mathbb{V}[\,\hat{\pi}_{ac}\,] = \mathbb{V}[\,P\,]$$

$$\frac{n\pi(1-\pi)}{(n+2)^2} = \frac{n\pi(1-\pi)}{n^2}$$

The numerators are the same, but the denominators differ. So, these two will never be equal. In fact, since the denominator of the Agresti-Coull estimator is larger, we know that $\hat{\pi}_{ac}$ will *always* have a lower variance than P. Thus, the Agresti-Coull estimator is more consistent than the sample proportion. As such, if all we care about is precision, we should use $\hat{\pi}_{ac}$ to estimate π.

2.2.3 Mean Squared Error

Bias speaks solely to large-scale behavior of the estimator. Variance speaks solely to the concentration of the estimator, to individual-scale behavior. An ideal estimator would have both low bias *and* high precision. In other words, it is most important to have the estimator concentrated (precision) about the population parameter (unbiased).

Estimators that are always close to the parameter are better than estimators that are much more spread out, even if the former is biased and the latter is unbiased. To evaluate the average closeness of the estimates, we have the mean squared error (MSE).

Definition 2.3. *Let P be an estimator of π. The **mean squared error** of P is defined as*

$$\text{MSE}\,[\,P\,] := \mathbb{E}\left[\,(P-\pi)^2\,\right]$$

I leave it as an exercise to prove that

$$\text{MSE}\,[\,P\,] = (\text{bias}[P])^2 + \mathbb{V}[\,P\,]$$

This second equation is useful as it shows the mean squared error combines accuracy and precision into a single value that measures both bias and variance.

Quick Note

Before continuing, it may be helpful to review the definitions of the expected value and the variance (Defns. 2.6 and 2.8 on page 39). A quick refresh will help to make the following a bit more meaningful.

Now, let us calculate the MSE of the sample proportion.

$$\text{MSE}[\,P\,] = (\text{bias}[P])^2 + \mathbb{V}[\,P\,]$$
$$= 0^2 + \mathbb{V}[\,P\,]$$
$$= \frac{\pi(1-\pi)}{n}$$

Since the sample proportion is unbiased for P, its MSE is just its variance.

The MSE of the Agresti-Coull

Ultimately, the mean squared error is used to *compare* estimators. For instance, knowing that the MSE is $\pi(1-\pi)/n$ tell us little. To make the most use of it, let us compare the MSE of the sample proportion to that of the Agresti-Coull estimator. This will allow us to determine which of the two estimators is more concentrated around π.

$$\text{MSE}[\,\hat{\pi}_{ac}\,] = \left(\frac{1-2\pi}{n+2}\right)^2 + \frac{n\pi(1-\pi)}{(n+2)^2}$$
$$= \frac{(1-2\pi)^2 + n\pi(1-\pi)}{(n+2)^2}$$
$$= \frac{(4-n)\pi^2 - (4-n)\pi + 1}{(n+2)^2}$$

To determine which of the two has a lower MSE, one can use analytical methods (set the two equal and "do math") or graphical methods. I leave the math as an exercise. It is not conceptually difficult, just a lot of algebra.

Alternatively, we can examine Figure 2.2 to see that when the population proportion π is close to 50% one should use the Agresti-Coull estimator. Just *how close* depends on the values of π and n.

One Estimator of π: The Sample Proportion 31

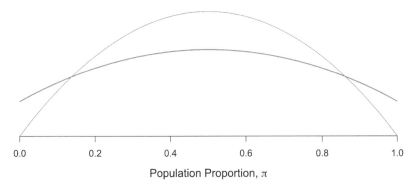

FIGURE 2.2
The mean squared error, as a function of π, for the sample proportion and the Agresti-Coull estimators ($n = 10$). Note that the Agresti-Coull estimator (thick curve) has a lower MSE for much of the domain of π. For the code to create this graphic, please see Section 2.6.3 (page 42).

Specifically, Figure 2.2 shows the MSE as a function of π when the sample size is $n = 10$. From that graphic, we see that the Agresti-Coull (thick) has a lower MSE than the sample proportion (thin curve) for values of π between 0.14 and 0.86, approximately.

This conclusion holds for other sample sizes. However, larger sample sizes has two effects: First, they decrease the benefit of using the Agresti-Coull estimator; the difference in MSE values converges to zero. Second, the region in which the Agresti-Coull estimator is superior shrinks.

2.2.4 Final Thoughts on Estimators

Everything in statistics is estimators. This is because nothing in applied statistics can be stated with mathematical certainty. Spending time comparing estimators in terms of bias, variance, and mean squared error encourages one to really come to a deeper understanding that the best estimators we have are fuzzy, that the very use of statistics forces us to admit everything is an estimate. Only providing the estimate is virtually worthless. One must *also* provide an estimate of how precise it is.

In the next section, I briefly introduce confidence intervals as a set of reasonable values for the population parameter arising from this natural variability. These confidence intervals help to enforce the understanding that a random sample leads to a random estimate. In Chapter 3, I spend an inordinate

2.2.5 Likely Sample Proportions

For the sample proportion estimator, the maximal variance (greatest uncertainty in the estimate) happens when $\pi = 0.500$. At this point, $\mathbb{V}[\,P\,] = 1/(4n)$ and the standard deviation is $\sigma = 1/(2\sqrt{n})$. Recall from your elementary statistics course that the Empirical Rule states that about 95% of the observations will fall within 2 standard deviations of the mean.[WDA14] Thus, we expect about 95% of the sample proportions to be within $2/(2\sqrt{n}) = 1/\sqrt{n}$ of the population proportion—the value we are trying to estimate.

That means about 95% of the sample proportions will be between

$$P_- = \pi - \frac{1}{\sqrt{n}}$$

and

$$P_+ = \pi + \frac{1}{\sqrt{n}}$$

For ease of writing, we can combine these two equations into

$$P_\pm = \pi \pm \frac{1}{\sqrt{n}}$$

Here, $P_\pm$ are the values of the two endpoints. The first (lower) is obtained by using the $-$; the second (upper), $+$.

Quick Example

Given the data from the Survation survey above, if the actual support for Scottish independence was 50%, then about 95% of the surveys would have sample proportions within $1/\sqrt{1061} = 3.07$pp of 50%; that is, between $0.5000 - 0.0307 = 0.4693$ and $0.5000 + 0.03070 = 0.5307$.

If the sample proportion is actually above 53.07% (or below 46.93%), then one would expect the reality that π is above (or below) 50%. That the Survation survey showed a support rate of 46.75% provides some evidence that $\pi < 0.500$.

2.2.6 Illustrating the Uncertainty

Because it is central to an understanding of why that uncertainty exists, let us perform a quick statistical experiment using random draws from a Binomial

One Estimator of π: The Sample Proportion

distribution. The following line produces a single vote for independence (one person), given that support is at 50%:

```
rbinom(n=1, size=1, prob=0.500)
```

When I ran this line in R, I got 0. This indicates that *my* single simulated voter voted against independence.[5]

If I would like to simulate 1061 such voters, perhaps simulating a single Survation-style poll, I would use

```
rbinom(n=1, size=1061, prob=0.500)
```

I got 514. That is, in this simulated poll, 514 people out of the 1061 asked said they supported independence. That is, this poll had $514/1061 = 48.4\%$ support independence.

Note: *Your answers will (most likely) differ from mine, since these are draws from a random population. This caution echoes the fact that two identical polls will likely differ in their results because they are drawing different random samples.*

To illustrate the variability in the polls, let us generate a large number of such polls and see how much the results vary. The following code generates 10,000 such polls, saves the results in the variable `pollResults`, creates a graphic of those results, and determines the middle 95% of poll results:

```
pollResults = rbinom(n=10000, size=1061, prob=0.500)
hist(pollResults)
quantile(pollResults, c(0.025,0.975))
```

While your results may differ, the conclusions should be quite similar. The histogram shows that there is a large variation in the outcomes (see Figure 2.3). The middle 95% of those results range from 499 to 562 (the dark area). Thus, if the support for independence really is 50%, we would not be surprised to get poll outcomes between 499 and 562, between 47.0% and 53.0%, between $50\% \pm 3\%$. If we observe outcomes outside that region, then we would have evidence that one of our assumptions is not correct. Usually, if the observed sample proportion is outside the central region, in the "rejection region," we claim we have evidence that $\pi \neq 0.500$.

---

[5] All calculations and simulations in this book are done using the R Statistical Environment.[R C18] This is introduced in Appendix A on page 282.

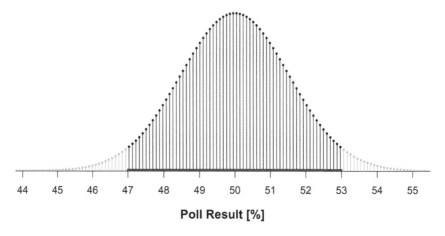

FIGURE 2.3
Histogram of sample poll results when $\pi = 0.500$ and $n = 1061$. Note that there is a very large spread (uncertainty) in the observed sample proportions. About 95% of the sample proportions are between 47% and 53%, the dark area.

2.2.7 The Three Assumptions

Note that the above is based on a few assumptions. First, it is based on assuming support for independence is $\pi = 50\%$. This is not a bad assumption if one only cares about which of the two sides will win. If the poll returns more than 562 in favor of independence, then there is statistically significant evidence that more than half of the population supports independence. Conversely, if fewer than 499 in the poll support independence, then there is statistically significant evidence that fewer than half support independence. In the Survation poll covered in the chapter introduction, there were 496 responses in favor of independence. Thus, there is evidence that the independence referendum will fail, that $\pi < 0.500$.

Second, it is based on the assumption that a person's probability of being selected is constant throughout the poll. This requirement can be met in two ways: first, that each individual can be contacted multiple times; second, that the population is infinite. Neither really reflects the reality of polling.

Third, it is based on the assumption that the probability of selecting any specific person is the same. This is problematic, as the contact method is frequently the telephone, and people have different exposures to telephones (from a large family served by a single landline to an individual served by a couple of smart phones and a landline).

2.3 SRS without Replacement

Thus far, we have examined a very simple sampling scheme, one in which each person in the population has the same likelihood of being contacted at each contact point. This means each person could be contacted multiple times. If this seems like a waste of effort and a loss of efficiency, then you may be right. The obvious simple improvement on this sampling scheme is to remove the person from the population once contacted.

Unfortunately, this small change forces us to change the distribution. The Binomial distribution requires that the probability of being contacted be constant throughout the data-gathering process. When we sample *without* replacement (as here), the probability changes from contact to contact in a specific manner. Because of this particular change, the correct distribution is the Hypergeometric distribution.[6]

From the discussion, it is clear that there are three pieces of information needed to specify the distribution: size of the population, size of the sample, and the proportion of successes in the population. Different sources will use different ways of specifying these three parameters, but they will need to use all three of them in some manner.

Because I am illustrating the similarities and differences between the Binomial and the Hypergeometric distributions here, I will use π as the proportion of successes in the population, n as the sample size, and N as the population size. This is not the usual parameterization, but it allows me to illustrate the main similarities and differences between the two distributions (Table 2.1).

Examine Table 2.1. First, note that the expected values are the same. This is helpful, as it allows us to use the estimators we created for the Binomial distribution. The sample proportion remains an unbiased estimator of π. Similarly, the Agresti-Coull estimator remains biased.

Second, note that the variance for the Hypergeometric distribution is smaller than for the Binomial by a factor of $\frac{N-n}{N-1}$. This shows the improvement in efficiency when one does not allow for asking the same person multiple

---

[6]If you have never heard of the Hypergeometric distribution, you are not alone. There is a definite improvement when using it. However, that improvement is incredibly minor when dealing with population sizes as low as 2000. Since most political divisions are larger than this, the advantages of using the Hypergeometric are drowned out by the increased math... as this section shows.

TABLE 2.1
Comparisons between the Binomial and the Hypergeometric distributions in terms of the expected value and the variance. Note that the expected values are the same, but the variance of the Hypergeometric is smaller.

Distribution	Expected Value	Variance
Hypergeometric	$n\pi$	$n\pi(1-\pi)\frac{N-n}{N-1}$
Binomial	$n\pi$	$n\pi(1-\pi)$

times. This improvement disappears as the sample size decreases and as the population size increases. There is no improvement when the sample size is $n = 1$ or when the population is infinite.[7]

Finally, note that the variance reduction (efficiency improvement) tends to be rather small when dealing with typical population and sample sizes in usual surveys. For instance, the population of Scotland is approximately 5.5 million. Thus, the efficiency improvement for the Survation survey introduced at the start of this chapter has a variance that is

$$\frac{5,500,000 - 1061}{5,500,000 - 1} \approx 0.999807$$

the size of one had they allowed for multiple contacts. In other words, if Survation asked the 1061 people allowing for asking the same person multiple times, the variance is equivalent to Survation asking 1060.8 people without allowing for repetition—the improvement in implementing this more-complicated scheme is minor.

Because of this, there is little reason to repeat the calculations of this chapter for this new sampling scheme. The Binomial distribution is a really good estimating distribution for when the sampling scheme is simple random sampling without replacement. That is, the extra precision is not worth the additional complexity.

2.4 Conclusion

In the next chapter, we will continue working with simple random sampling. However, we will spend our time learning about how to properly indicate the level of precision in an estimate. This leads to **confidence intervals**

---

[7]Check that these results make sense. The Hypergeometric is equivalent to the Binomial distribution when either $n = 1$ or $N = \infty$.

and **credible intervals**. The first may be familiar to those who have taken introductory statistics recently. The latter, probably not.

In Chapter 4, we will move away from the conceptually simpler simple random sampling to the more frequently used stratified sampling. The concepts will remain the same. The mathematics will become more complicated. The purpose of this chapter was to introduce the reader to the concepts of polling. These included sampling schemes and estimators. I also gave a tantalizing hint about confidence intervals.

This chapter used the simplest of sampling schemes: simple random sampling. This decision allowed us to dip our toes into the richness of the mathematics underlying polling. Simple random sampling led us to the Binomial distribution and the sample proportion as the natural estimator. From there, we looked at estimating π based on a single sample.

One interesting estimator for the population proportion arising from a Bayesian understanding is the Agresti-Coull estimator. Here, we examined it in detail, comparing it to the usual sample proportion estimator. It surpassed the sample proportion estimator in all areas except for bias.

Again, there is much more to polling. This chapter examined the *simplest of cases*, allowing us to better understand what the polls are really telling us—and what they **cannot**.

2.5 Extensions

In this section, I provide some practice work on the concepts and skills of this chapter. Some of these have answers, while others do not.

Review

1. Why is the Agresti-Coull estimator not often used in polling analysis?
2. How would you convince a person that an estimator with a lower MSE is preferred to one that is unbiased?
3. Why is the conservative margin of error approximately $1/\sqrt{n}$?
4. What does the thick line at the bottom of Figure 2.3, page 34, represent?

Conceptual Extensions

1. There were four counties (local authorities) in which the 2014 Scottish independence referendum received more than 50% of the votes cast (Figure 2.1). What do they have in common?

2. The turnout for the 2014 Scottish independence referendum was not 100%. Assuming that those who did not vote in each county were similar to those who did, would a higher turnout have helped independence or not?

3. The Survation poll estimated independence support to be 47%, which was 2.3% too high. Was this poll "too far off"? This question gets at your understanding of the confidence interval.

4. Using words, answer these two questions:

    (a) How do we know that the Binomial distribution is equivalent to the Hypergeometric distribution when $n = 1$?

    (b) How do we know that the Binomial distribution is equivalent to the Hypergeometric distribution when $N = \infty$?

5. Using a mathematical proofs, answer these two questions:

    (a) How do we know that the Binomial distribution is equivalent to the Hypergeometric distribution when $n = 1$?

    (b) How do we know that the Binomial distribution is equivalent to the Hypergeometric distribution when $N = \infty$?

Computational Extensions

1. Using Defn. 2.3, prove $\text{MSE}[\,P\,] = \text{bias}^2[P] + \mathbb{V}[\,P\,]$.

2. Repeat all of the Survation calculations for this chapter for a poll in which 256 out of 500 people supported Scottish independence. Do the conceptual conclusions significantly differ?

2.6 Chapter Appendix

This appendix contains some "review" material on random variables and their mean (expected value) and variance. It also covers the R functions mentioned in this chapter.

2.6.1 Some Useful Definitions

The following are some definitions you may find useful in your unending quest to better understand statistics and polling.

Random Variable

This section defines a random variable. The first definition is mathematically phrased. The second definition is actually helpful for us.

Definition 2.4. *A **random variable** is a function mapping a sample space to a measurable space; that is, $X : \Omega \mapsto S$. Note that Ω must be the sample space of a **probability triple** $(\Omega, \mathcal{F}, P)$, where $\mathcal{F}$ is a set of events and P is a probability function.*

While this definition is very helpful for studying the mathematical underpinnings of probability theory, it is definitely overkill for us. As such, I prefer the following "definition."

Definition 2.5. *A **random variable** is the outcome of a future experiment (or observation).*

This definition focuses on the fact that the random variable cannot be known until the experiment or the observation is performed. This emphasizes the randomness of the variable because the future is inherently uncertain.

Expected Value

The expected value is the long-run average of experimental outcomes. It says nothing about the likelihood of any specific outcome.

Definition 2.6. *Let Y be a random variable. The **expected value** of Y, symbolized by μ, μ_y, or $\mathbb{E}[\,Y\,]$, is defined as the long-run average of Y.*

In the discrete case, its formula is

$$\mathbb{E}[\,Y\,] := \sum_{y \in S} y\, \mathbb{P}[\,Y = y\,]$$

Here, S is the sample space, the set of all possible values of Y.

With this definition, it is elementary to prove that the expected value is a linear operator.

Theorem 2.7. *Given Y is a random variable with a and b both non-random, $\mathbb{E}[\,aY + b\,] = a\,\mathbb{E}[\,Y\,] + b$.*

Proof. This proof follows from the definition of the expected value:

$$\begin{aligned}
\mathbb{E}[\,aY+b\,] &= \sum_{y \in S} (aY+b)\,\mathbb{P}[\,Y=y\,] \\
&= \sum_{y \in S} aY\,\mathbb{P}[\,Y=y\,] + \sum_{y \in S} b\,\mathbb{P}[\,Y=y\,] \\
&= a \sum_{y \in S} Y\,\mathbb{P}[\,Y=y\,] + b \sum_{y \in S} \mathbb{P}[\,Y=y\,] \\
&= a\,\mathbb{E}[\,Y\,] + b\,1
\end{aligned}$$

Thus, the conclusion follows that $\mathbb{E}[\,aY+b\,] = a\,\mathbb{E}[\,Y\,] + b$.

∎

Variance

The variance is a measure of the uncertainty in the outcomes of a random variable. Here is its definition:

Definition 2.8. *The **variance of a discrete random variable** Y is an average distance between the values and the center. It is symbolized as σ^2, σ_y^2, or $\mathbb{V}[\,Y\,]$. Mathematically, its formula is*

$$\mathbb{V}[\,Y\,] := \sum_{y \in S} (y-\mu)^2\,\mathbb{P}[\,Y=y\,]$$

With this definition, it is elementary to prove that the variance is a quadratic operator.

Theorem 2.9. *Given Y is a random variable with a and b both non-random, $\mathbb{V}[\,aY+b\,] = a^2\,\mathbb{V}[\,Y\,]$.*

Proof. This proof follows from the definition of the variance:

$$\begin{aligned}
\mathbb{V}[\,aY+b\,] &= \sum_{y \in S} \big((aY+b) - \mathbb{E}[\,aY+b\,]\big)^2\,\mathbb{P}[\,Y=y\,] \\
&= \sum_{y \in S} \big((aY+b) - (a\mu+b)\big)^2\,\mathbb{P}[\,Y=y\,] \\
&= \sum_{y \in S} \big((aY - a\mu)\big)^2\,\mathbb{P}[\,Y=y\,] \\
&= \sum_{y \in S} \big(a(Y-\mu)\big)^2\,\mathbb{P}[\,Y=y\,]
\end{aligned}$$

$$= \sum_{y \in S} a^2 \big((Y - \mu)\big)^2 \, \mathbb{P}[\, Y = y\,]$$

$$= a^2 \sum_{y \in S} (Y - \mu)^2 \, \mathbb{P}[\, Y = y\,]$$

Thus, the conclusion follows that $\mathbb{V}[\, aY + b\,] = a^2 \, \mathbb{V}[\, Y\,]$. ∎

Continuous Distributions

The above definitions are written for discrete random variables like the Binomial distribution. Similar definitions hold for continuous distributions like the Beta. Just change the summation $\sum_S$ to an integration $\int_S$ and the probability function $\mathbb{P}[\, Y = y\,]$ to the probability density function $f(x)$.

2.6.2 R Functions Used

A statistical program is invaluable in gaining a better understanding of statistics and its/their relationships to the world around us. The following R functions were featured in this chapter.

- `rbinom`

  All basic distributions are represented in R. For each, there are usually four things that can be calculated on the distribution: likelihood, cumulative probability, quantile, and a random value. The structure of R probability functions is a prefix, denoting what needs to be calculated, and a stem, denoting which distribution. Here, `binom` indicates a Binomial distribution is being used and `r` indicates random values are being generated. For instance, `rbinom(n=10000, size=1061, prob=0.500)` generates 10,000 random values drawn from a $\text{BINOM}(n = 1061, \pi = 0.500)$ distribution.

- `quantile`

  This calculates the specified quantile(s) of a vector of numeric values. For instance, `quantile( pollResults, c(0.025,0.975) )` calculates the 2.5 and 97.5th percentiles of the values in the `pollResults` variable.

- `hist`

  This produces a basic histogram of the provided numeric variable. As with all R graphics functions, there are enough options to create just about *any* histogram your heart desires.

2.6.3 A Bonus Graphic

In Section 2.2.3 (page 31), I provided a graphic showing the difference between the MSE of the sample proportion estimator and of the Agresti-Coull estimator. Here is code you can use to produce it. I am providing it here in case you are interested in seeing the benefits of the Agresti-Coull estimator for different sample sizes, n.

```
### Agresti-Coull vs. Sample Proportion

n = 10

xx = seq(0,1,length=1e4)
ySP = xx*(1-xx)/n
yAC = ( (4-n)*xx^2 - (4-n)*xx + 1 )/( (n+2)^2 )

par( mar=c(3,0,0,0)+1 )
par( cex.lab=1.2, font.lab=2 )
par( xaxs="i", yaxs="i" )

plot.new()
plot.window( xlim=c(0,1), ylim=c(0,0.03) )

axis(1)
title( xlab=expression(paste("Population Proportion, ",pi)),
    line=3)

lines(xx,ySP, lwd=1, col="green4")
lines(xx,yAC, lwd=2, col="red1")
```

3

Interval Estimation

In the previous chapter, we were introduced to simple random sampling (SRS). Even though polling firms tend to not use this sampling scheme, it is mathematically clean and clear. We can leverage it to learn more about statistics and how they can give us some insight into elections and polls.

Last chapter, we also looked as estimators and estimates. Such terms as bias, variance, and mean squared error were introduced. This chapter looks at intervals. These represent the uncertainty in our estimates. Without some measure of uncertainty, an estimated support level of 42% is meaningless.

Kingdom of Bhutan: The General Election of 2024
I found out that one of my students is going to spend a large part of next year in Bhutan as a part of her ecological research. I envy her so much because Bhutan is on the list of countries I would like to visit.

Why Bhutan? Well, Bhutan is a small Himalayan country landlocked between China and India. It is very beautiful and the home of the tallest mountain in the world that has yet to be climbed, Gangkhar Puensum, which is Dzongkha for "White Peak of the Three Spiritual Brothers." Most importantly, Bhutan is a biodiversity hot spot, which is why my student is spending time there next year.

Politically, I also find it quite interesting. In the early 2000s, King Jigme Singye Wangchuck pushed for Bhutan to move from an absolute monarchy to a constitutional monarchy. As a result of this action, the committee he selected presented him with the 2005 Constitution. Two years later, after preparing the citizens of Bhutan for the transition to democracy, the first parliamentary elections were held.

Since 2007, Bhutan has held elections for the National Assembly (lower house of parliament) every five years using their rather interesting electoral

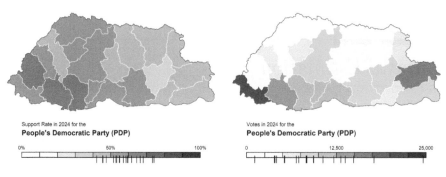

FIGURE 3.1
Result maps of the 2024 second-round general election in Bhutan. The left map illustrates the proportion of votes in favor of the People's Democratic Party by dzongkhag. The right map illustrates the number of votes received by the PDP. These data are from the Election Commission of Bhutan.[Ele24]

system. The National Assembly consists of 47 members elected from single-member districts. They are elected in a two-round system. The first round has candidates from the several parties stand for election. However, the vote counts are aggregated across the country for the sole purpose of determining which two parties will stand in the second round. The second round has candidates from just the two top parties stand for election in the divisions.

Thus, it is quite possible that a candidate will win the division in the first round but not be able to run in the second round because their party did not receive enough votes in the kingdom. For instance, in the 2023–24 election, Druk Phuensum Tshogpa (DPT) won the most votes in Khar Yurung, but could not stand for the second-round (which was won by Sangay Thinley of the Bhutan Tendrel Party).[Ele24]

It seems to me that there is a major benefit to this electoral system: Parties with limited regional support will not be elected to the parliament. This keeps the National Assembly more focused on national concerns.

In the 2023–24 election, the People's Democratic Party (PDP) was very popular with the people. There was no question that they would be in the second round. However, the second-place party was a toss-up between the Bhutan Tendrel Party (BTP), the Druk Nyamrup Tshogpa (DNT) party, and the Druk Phuensum Tshogpa (DPT) party. Once the votes were counted, it was the Bhutan Tendrel Party that moved on to the second round with a bit under 20% of the votes cast.[Ele24] Thus, the second round was between candidates from the People's Democratic Party and the Bhutan Tendrel Party.

Figure 3.1 provides two maps. The one on the left shows the proportion of the vote going to People's Democratic Party (PDP) in each of the

Margin of Error 45

20 dzongkhags. This shows that the PDP has a rather consistent support across Bhutan, with unsurprising variations.

The map on the right shows the *number of votes* received by the PDP in each dzongkhag. According to this map, the PDP received about 25,000 votes in the Samtse dzongkhag (darkest region on the bottom left).

These are two maps of the same data. And yet, they tell complementing stories of the election. What stories do they tell separately, and what story do they tell together?

Overview

In the previous chapter, we introduced some of the mathematics underlying simple random sampling. By the end of last chapter, you were able to use raw polling results to estimate the support level for a candidate. This *point estimate* is a single number. It provides the most likely outcome.

However, there is important information in the sample that we have entirely ignored to this point: precision. While we have discussed precision previously (Section 2.2.2), even running a short experiment to illustrate that sample proportions are random variables (Section 2.2.5), we have not covered how to provide that precision in a reported support level.

This chapter looks at three ways of calculating and presenting precision in our reported numbers. The first two are suitable for frequentists; the third, for Bayesians.

3.1 Margin of Error

The fundamental purpose of providing estimates of precision is to convey to the reader how uncertain one is in the estimate. Thus, saying that a candidate has 45% support does not tell the entire story. More information must be provided to give a truer story about the support level, the level of precision of the estimate.

One way to provide that additional information is to use the **margin of error** (MOE). In general, the margin of error is expressed as a percentage of the sample statistic or an absolute value in terms of units of measurement. For example, if a survey estimates that 52% of the respondents support Candidate

X, the margin of error might be reported as ±3 percentage points at a 95% confidence level.

By reporting the MOE along with the sample statistic, journalists and researchers can provide a more complete picture of the uncertainty surrounding their estimates and help readers understand the potential range of values that the true population parameter might take. However, it is important to note that the MOE only provides a *lower bound* on the total error. Furthermore, the margin of error only describes the amount of variation in the estimate due to the sampling method used. It **does not** provide estimates of uncertainty due to *other* sources of error, such as the sample being unrepresentative.

3.1.1 Variability in the Sample

Back in Section 2.2.6, we did a short experiment that showed how much the sample proportions varied from sample to sample. This variation was due to the fact that the sample was random. Since the sample is random, everything calculated based on that sample is also random. Specifically, the sample proportion is a random variable.

In that experiment, we saw that 95% of the poll results had between 499 and 562 successes. Thus, if we observed fewer than 499 successes, we would conclude that the population support was less than 50%. The logic of this experiment was:

- Assume: $\pi = 0.500$
- Assume: $n = 1061$
- Assume: the sample is representative
- Conclude: We should observe between 499 and 562 successes

This logic follows the structure of a classic syllogism. Note that there is one bit of "fuzziness" in the otherwise perfect logical form: "should." Because we are working with randomness, we cannot conclude that we *will* observe between 499 and 562 successes, only that we *should*. There is a non-zero probability of observing fewer than 499 or more than 562 successes. However, we deem those "unlikely at the 95% confidence level."

Thus, using contrapositive logic, if we *do not* observe between 499 and 562 successes, then at least one of the assumptions is incorrect... or we are unlucky. This logic was the source of Section 2.2.7.

3.1.2 The Formula

Combining the above into one coherent whole, we can conclude the following series:

$$X \sim \text{BINOM}(n, \pi)$$

Using the Central Limit Theorem, this means

$$X \mathrel{\dot\sim} \text{NORM}(n\pi,\ n\pi(1-\pi))$$

The "$\dot\sim$" indicates that X only *approximately* follows this distribution (that is what the dot on top of the tilde indicates: "approximate"). The approximation improves as the sample size increases. Now, since we eventually want to use P in our formula instead of X, we divide by n:

$$P = \frac{X}{n} \mathrel{\dot\sim} \text{NORM}\left(\pi,\ \frac{\pi(1-\pi)}{n}\right) \tag{3.1}$$

While this will work for us, it is usual to perform a z-transformation to ensure that the distribution follows a Z distribution, a.k.a. a "standard Normal" distribution.

$$\frac{P - \pi}{\sqrt{\pi(1-\pi)/n}} \mathrel{\dot\sim} \text{NORM}(0,1)$$

This quantity on the left is frequently referred to just as "Z" in this context. It is a test statistic that is used to test hypotheses about population proportions.

Since this quantity follows a standard Normal distribution, we know that it will be between −1.96 and +1.96 about 95% of the time. This is a consequence of the distribution being standard Normal (Figure 3.2). With that, we can obtain reasonable bounds on P if we know the value of π or on π if we know P. The first led to the results in Section 2.2.5. The second is useful if we are estimating π, which is what we usually do.

The following proves the formula for reasonable bounds on π:

$$\frac{P - \pi}{\sqrt{\pi(1-\pi)/n}} \mathrel{\dot\sim} \text{NORM}(0,1) \tag{3.2}$$

The first bound arises when the quantity on the left is set to +1.96:

$$\frac{p - \pi}{\sqrt{\pi(1-\pi)/n}} = +1.96$$

$$p - \pi = +1.96\sqrt{\frac{\pi(1-\pi)}{n}}$$

$$-\pi = -p + 1.96\sqrt{\frac{\pi(1-\pi)}{n}}$$

$$\pi = p - 1.96\sqrt{\frac{\pi(1-\pi)}{n}}$$

Thus, the lower bound for reasonable values of π is $p - 1.96\sqrt{\pi(1-\pi)/n}$. I leave it as an exercise for you to show that the upper bound for reasonable values of π is $p + 1.96\sqrt{\pi(1-\pi)/n}$. Note that we are adding/subtracting the same number to the sample proportion, p. That is, the endpoints of reasonable values for π are

$$\pi = p \pm 1.96\sqrt{\frac{\pi(1-\pi)}{n}} \tag{3.3}$$

We shall name the quantity to the right of the $\pm$ the **"margin of error"** because that single number estimates the reasonable amount of error due to sampling.

Thus, for the 95% confidence level, the formula for the margin of error is

$$E = 1.96\sqrt{\frac{\pi(1-\pi)}{n}} \tag{3.4}$$

Note that different confidence levels will lead to different coefficients. For instance, a 99% confidence level changes the 1.96 to 2.576; a 90% confidence level changes it to 1.645. However, since we tend to default to a 95% confidence level, I am happy to stay with 1.96.

The Problem with π

Normally, I would provide a quick example to see the above in action. However, we have a problem. Eqn. 3.4 has the unknown variable π on the right side of the formula. Thus, apparently, one can not calculate the margin of error without knowing π. But, if we *knew* π, then we would not need to poll. Quite the conundrum.

There are (at least) two simple solutions to this problem. And, in true statistics fashion, they give different results. The first solution is to substitute 0.500 for π in the formula. This tends to give wider margins of error, which allows the researcher to provide *conservative* estimates of the support in the population. This should be used if $\pi = 0.500$ is a reasonable value for π, given the problem before us.

The second solution is to replace π with p; that is, use the sample proportion instead of π. This was Abraham Wald's solution to the problem. Unfortunately, this will bias the margin of error in the direction of how p differs from π. If, however, the sample is a good sample, then this bias tends to be negligible. If the sample is not a "good sample" (is not representative of the population), then it may give very bad estimates.

Margin of Error 49

Now for a Quick Example

Let us pretend that Knox College (TKS Polling) conducted a poll in Bhutan asked $n = 500$ people about their party preference. Of those asked, $x = 86$ stated their preference for the Bhutan Tendrel Party. From this poll, we calculate the point estimate to be $p = 86/500 = 17.2\%$.

Using the formula above (Eqn. 3.4), and using $\pi = 0.500$, the margin of error is 4.38 percentage points. Using the Wald interval, the margin of error is 3.31pp. Thus, we would report the poll results as either

$$17.2\% \pm 4.38$$

or as

$$17.2\% \pm 3.31$$

As expected, the conservative interval is wider than the Wald interval.

Note that these are not the only two ways of dealing with the π in the formula. There are a few others, including the Agresti-Coull and something called the credible interval. However, I think I have covered margins of error sufficiently to give you some insight into the parts, what affects it, and what should be declared when stating them.

3.1.3 Why 1.96? When not 1.96?

One number that arises in statistics over and over again is 1.96. This number arises from the fact that the random quantity in Eqn. 3.2 follows a standard Normal distribution (Figure 3.2). The fact that it does means that we will observe values in the unshaded region 95% of the time simply due to randomness in the sampling method. It is a fact of life that cannot be changed. The −1.96 and +1.96 separate the unshaded region from the shaded regions (the tails).

Thus, it arises from two sources: Simple randomness and our decision to calculate a 95% interval. Had we decided on a different confidence level, the endpoints would have to change. To ensure only 90% is in the unshaded region, the endpoints would have to change to ± 1.645; to ensure that 99% is in the unshaded region, ± 2.576.

Why might one want to change the confidence level? In other disciplines, it may be useful. For instance, particle physics apparently uses a confidence level of 99.9999% when determining if they have detected a new particle (they want to make doubly sure the resulting hullabaloo is warranted). Electoral forensics uses a confidence level of 99.9% in testing for unfairness (you don't

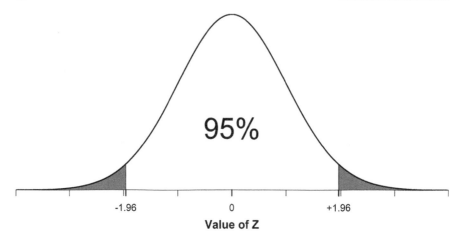

FIGURE 3.2
An illustration of the origin of the 1.96 in the formulas. Observations occur in the *un*shaded area 95% of the time. This region is separated from the shaded region by the values −1.96 and 1.96.

want to claim a government is lying unnecessarily). For estimating candidate support, I see no reason to use a confidence level other than 95%.

3.2 Confidence Intervals

In the previous section, I jumped right to the margin of error. Recall that the margin of error provides an estimate of the variability caused by the sampling method. Thus, in the Bhutan quick example, the 3.31 is an estimate of the reasonable variability due to simple random sampling with this particular sample size. Thus, the strength of using the margin of error is that it forces us to recognize that variability.

However, one will frequently want to provide an "*interval of reasonable values*" for the population proportion, π. That is, we may want to focus on the uncertainty in the estimate. If that is our goal, then we would provide a "**confidence interval**" instead of a margin of error.

3.2.1 The Formulas for Confidence Intervals

Given the construction in Eqn. 3.3, it is *reasonable* to claim that π is between

$$\pi_l = p - 1.96 \sqrt{\frac{\pi(1-\pi)}{n}}$$

and

$$\pi_u = p + 1.96 \sqrt{\frac{\pi(1-\pi)}{n}}$$

In this context, what do the values of π_l and π_u represent? These are the points defining the boundary between "reasonable" and "unreasonable" values of π. If the claimed value of π is beyond these two values, then there is sufficient evidence that the claimed value is *not* reasonable, given this data.

This interpretation came from the margin of error discussion above. The only thing we will add to the discussion is how we represent this information. The confidence interval is all values between π_l and π_u.

We can, of course, write the above two equations in the more compact form we are familiar with:

$$\pi = p \pm 1.96 \sqrt{\frac{\pi(1-\pi)}{n}} \qquad (3.5)$$

Again, change the 1.96 for whichever confidence level coefficient you wish (e.g., 2.576 for 99%; 1.645 for 90%).

Quick Example

Given the results of the Bhutanese TKS Polling survey, and using the conservative confidence interval, we can estimate the proportion of voters supporting the Bhutan Tendrel Party to be between

$$17.2\% - 4.38 = 12.82\%$$

and

$$17.2\% + 4.38 = 21.58\%$$

with 95% confidence.

Thus, claiming that π is any value in the interval (12.82% to 21.58%) is reasonable, given the survey results and our definition of reasonable. For instance, it would be *unreasonable* to claim that the Bhutan Tendrel Party was winning the election (i.e., that $\pi > 50\%$) based on these results at the 95% level.

More on the Trouble with π

Note again that Eqns. 3.3 and 3.5 both have the variable π on both sides of the equation. In the section on the margin of error, we saw two relatively easy solutions to the problem of a π on both sides (use 0.50 for π and use p for π). And, in true statistics fashion, they gave slightly different results.

If we are working with confidence intervals, then there is a third solution. It is to just solve Eqn. 3.5 for π. This gives

$$\pi = \frac{\left(2np + Z^2_{\alpha/2}\right) \pm \sqrt{-4np^2 Z^2_{\alpha/2} + 4np Z^2_{\alpha/2} + Z^4_{\alpha/2}}}{2\left(n + Z^2_{\alpha/2}\right)} \quad (3.6)$$

Here, p is the sample proportion, n is the sample size, and $Z_{\alpha/2}$ is that distribution coefficient (1.96 if we have a confidence level of 95%; 2.576 for 99%; 1.645 for 90%, etc.).

Quick Example

Given the results of the Bhutanese TKS Polling survey, and using the Eqn. 3.6 confidence interval, we estimate the proportion of voters supporting the Bhutan Tendrel Party to be from 14.15% to 20.75% at the 95% confidence level.

Note that the above confidence interval (Eqn. 3.6) cannot be written in the form of $p \pm E$ as can the others (p is the sample proportion). That is because in solving for π, the center of the confidence interval shifted a bit. This is not a problem if this confidence interval is actually better. What do we mean by "better"? We mean that it contains the population proportion the stated percent of the time (95%, usually).

We will look at coverage shortly. However, before we do, let us see an example and look at the *actual* meaning of confidence intervals.

An Example of Confidence Intervals

Let us return to the Bhutanese survey. In that survey of 500 voters, let us say that the People's Democratic Party received 248 supporters. The following provides the calculated endpoints of 95% confidence intervals for this survey:

Method	Lower Bound	Upper Bound
Eqn. 3.5, 0.500 for π	0.4522	0.5398
Eqn. 3.5, p for π	0.4522	0.5398
Eqn. 3.6	0.4524	0.5397

Confidence Intervals

Note that the three estimates are quite similar to each other. This is because n is large and the support is near $\pi = 0.50$. Were we working with π closer to 0 or 1, the three estimators would markedly differ, as the estimated confidence intervals for the Bhutan Tendrel Party suggests:

Method	Lower Bound	Upper Bound
Eqn. 3.5, 0.500 for π	0.1282	0.2158
Eqn. 3.5, p for π	0.1389	0.2051
Eqn. 3.6	0.1415	0.2075

These confidence intervals have very different endpoints. Which of the three is best? That is: Which should we use?

3.2.2 Confidence Interval Performance

Confidence intervals are rather straight-forward to calculate—and to interpret. These are the intervals almost universally covered in an introductory statistics course. However, there is one very large weakness to using confidence intervals: They have no probabilistic meaning on a single experiment (poll). The closest thing to a "probability statement" that can be made is

> If we repeat the poll an infinite number of times, then 95% of the confidence intervals will contain π.

That is all.

However, that is more than enough to create a method for determining the quality of a confidence interval calculation method. The procedure comes from the statement:

1. Generate the outcome of a poll
2. Calculate the endpoints of a confidence interval based on that data
3. Determine if the interval contains (covers) the true value of π

This procedure will determine if *one* randomly selected interval covers the value of π. We just need to repeat the experiment an *infinite number of times* to determine the true coverage rate (or at least a very large number of times). Coverages closer to 95% are preferred.

Our Three Methods

In this chapter, we covered three methods to calculate the endpoints of a confidence interval. So, which is the "right" one to use?

As with most things in statistics, the answer depends on the particular parameters of the sampling. For instance, a sample of size $n = 500$ and $\pi = 0.20$ may have a different "better confidence interval" than one in which $n = 500$ and $\pi = 0.40$. Furthermore, while statistics can give *its* answer on which is best, it is also up to the *journalist* as to which can be more easily explained to the audience.

And so, from a statistics standpoint, to determine which confidence interval will be best, we will follow the steps laid out above.

Experiment: To determine which of the three methods is best for estimating the support level for the Bhutan Tendrel Party (assuming $\pi = 0.20$) and the People's Democratic Party (assuming $\pi = 0.40$) from the TKS Poll. The following give the coverage for a million polling samples.[1]

Method	Coverage for BTP	Coverage for PDP
Eqn. 3.5, 0.500 for π	0.984	0.950
Eqn. 3.5, p for π	0.949	0.950
Eqn. 3.6	0.950	0.950

The results are quite interesting. The three methods differ significantly when the proportion being estimated is far from $\pi = 0.50$, but those differences seem to vanish as π gets closer to 0.50.

Furthermore, there is little difference in coverage between the Wald method and the overly complicated equation (Eqn. 3.6). This seems to be true when the sample size is on the order of what is actually performed in real life ($n \geq 400$).

◇ ◇ ◇

---

[1] The code for this is given in the chapter's appendix (Section 3.6.2). As always, I encourage you to work with it to see how the computer can help *you* better understand the statistics of polling.

Credible Intervals 55

3.2.3 The Meaning of Confidence Intervals

Recall that the only probability statement that can be made using a confidence interval is

> If we repeat the poll an infinite number of times, then 95% of the confidence intervals will contain π.

Think about the consequences for a moment. That first clause is very important. If we perform the poll only once, then we cannot state the probability of π being in the interval is 95% (or 99% or whatever). We can *only* make some vague claim about how "confident" we are on its value. Section 3.6.2 (page 64) explores this in more detail.

A second consequence of this is that the endpoints of this particular confidence interval are not even involved in the probability statement. Personally, confidence intervals would be more easily understood—and used—if we could make probability statements using them. For instance, it would be really nice to be able to state something like this:

> The probability that the Bhutanese People's Democratic Party (PDP) wins the 2023 general election is 38%.

While it would be helpful to make some sort of probability claim about the value of π, it *cannot* be done using confidence intervals. To make such a claim, one needs to use Bayesian statistics, which I introduce in the next section and cover in depth in Chapter 5.

3.3 Credible Intervals

The confidence interval is based on the randomness of the observations when the population proportion is fixed. We could also attack the problem of estimating uncertainty by assuming the population proportion is unknown and either follows a probability distribution *or* our uncertainty about it implies a probability distribution. In taking this tack, we enter the realm of Bayesian statistics. Since this topic is rarely covered in introductory statistics, I merely introduce credible intervals here and punt the full discussion of Bayesian analysis to Chapter 5.

Credible intervals are a Bayesian analog of confidence intervals. They provide the range within which the true value of a parameter is likely to fall. Unlike confidence intervals, credible intervals represent the researcher's belief about the population proportion π based on prior information and the observed data.

They are constructed by finding the interval that contains a certain percentage (e.g., 95%) of the posterior probability for a given parameter. For example, if a Bayesian analysis produces a posterior distribution for the population support for the PDP party, a 95% **credible interval** would be the range that contains 95% of the *probability* for this distribution. Most importantly: This means that there is a 95% probability that the true PDP support falls within this interval, given the observed data and prior information.

Compared to confidence intervals, credible intervals have several advantages in Bayesian analysis. First, they directly represent the researcher's degree of belief about the parameter, rather than relying on frequentist concepts such as coverage rate or sampling distribution. Second, they can be used to compare different models or prior distributions and assess their impact on the posterior estimates. Third, they can be easily interpreted and communicated in terms of probabilities, making them more intuitive for many researchers and practitioners. Finally, they are flexible enough to work with non-probability sampling (which is beyond the scope of this book).

However, please note that credible intervals are *not* interchangeable with confidence intervals and should *not* be used as a direct replacement in frequentist analyses, no matter how wonderful that would be. They represent different concepts and methods of statistical inference and require different assumptions and calculations.

3.3.1 Bayes' Law

While I reserve the full treatment of Bayesian analysis for Chapter 5, I do need to introduce some of the terms and concepts here. Bayesian analysis arises from Bayes' Law.[Bay63] If we let π be the population proportion and x be the data, then Bayes' Law for estimating π can be written as

$$g(\pi \mid x) = \frac{\mathcal{L}(x \mid \pi) \, f(\pi)}{\int_0^1 \mathcal{L}(x \mid \pi) \, f(\pi) \, \mathrm{d}\pi}$$

In this equation,

- $g(\pi \mid x)$ is the probability distribution of π, given the new data;
- $\mathcal{L}(x \mid \pi)$ is the likelihood of the data, given the parameter;
- $f(\pi)$ is the initial (assumed) probability distribution of the parameter; and
- $\int_0^1 \mathcal{L}(x \mid \pi) f(\pi) \, \mathrm{d}\pi$ is a normalizing constant.

Credible Intervals

Because the denominator is uninteresting (it is required to ensure the left-hand side is a probability distribution), we frequently write Bayes' Law as

$$g(\pi \mid x) \propto \mathcal{L}(x \mid \pi) \, f(\pi)$$

The symbol $\propto$ means "is proportional to." When using this form, we need to ensure that the left side is a probability distribution. Frequently, we will do that simply by noticing that the product on the right "has the form of" a known distribution.

The distribution $f(\pi)$ is called the "**prior distribution**" of the parameter, because it is its assumed distribution *prior to* collecting the data, x. The quantity $\mathcal{L}(x \mid \pi)$ is called the "**likelihood function**" of the data. It is the probability of observing this particular data, given the current parameter values. The distribution $g(\pi \mid x)$ is called the "**posterior distribution**" of the parameter; it is the prior distribution updated with the new observations.

Bayesian analysis requires the analyst to make an assumption about the parameter's distribution before collecting data. This prior distribution should properly reflect current knowledge (expectations) about that parameter. For instance, the prior distribution for PDP support with no previous data will be very different than the prior distribution for PDP support where the last 100 polls all conclude $\pi = 0.40$. In the first case, a standard uniform distribution will be appropriate; it only imposes the requirement that π is between 0 and 1. In the second case, a Beta distribution with parameters $a = 300, b = 450$ may be appropriate; it states that the analyst expects π to be $40\% \pm 3.6$pp.

The Beta Distribution

Trust me on this. You should encode your prior knowledge and understanding of π using the Beta distribution. It just makes the calculations so much easier. Here is why: If the prior distribution is written as $\pi \sim \text{BETA}(a, b)$ and if you observe x successes and $n - x$ failures, then the posterior distribution will be

$$\pi \mid x \sim \text{BETA}(a + x, \; b + n - x)$$

Thus, the posterior distribution is just the prior updated with the observed number of successes (x) and failures ($n - x$). Believe me until you get to Section 5.3, where I prove this.

Thus, the hard part will be determining a and b in the prior distribution. If you have no prior information about the support level, then you should use $a = 1$ and $b = 1$. This is equivalent to the standard Uniform distribution.

However, if you do have information on the support level, the following mathematical function will help you calculate the appropriate values of a and b for your prior distribution.

Some "Helpful" Formulas

Let $\pi \sim \text{BETA}(a, b)$. Then, we know that the expected value of π is

$$\mu = \frac{a}{a+b}$$

and the standard deviation is

$$\sigma = \sqrt{\frac{ab}{(a+b)^2(a+b+1)}}$$

Simulatneously solving these two equations for a and b gives us the hoped-for formulas:

$$a = \left(\frac{\mu(1-\mu)}{\sigma^2} - 1\right)\mu$$

$$b = \left(\frac{\mu(1-\mu)}{\sigma^2} - 1\right)(1-\mu)$$

If you prefer to use the computer, this R function will calculate a and b for you given the expected value and standard deviation of π that you believe.

`betaPrior(mu, sigma)`

Illustrative Bayesian Example 1

As a "lengthy" example, let us do Bayesian analysis on the synthetic TKS Polling survey of party support in the 2023 first round of the Bhutanese general election.

To do Bayesian analysis, the first step is to specify a prior distribution on π. For this first example, let us assume we have no prior information about support for Bhutan Tendrel Party (BTP); that is, let us use the $\text{BETA}(1,1)$ distribution (a.k.a. the standard Uniform distribution). With this, and the observed values of $x = 86$ and $n = 500$, the posterior distribution is the $\text{BETA}(1+86,\ 1+500-86) = \text{BETA}(87,\ 415)$ distribution.

That's the "hard" part.

Now, you have enough information to calculate any and all probabilities you want. I suppose this is the "harder" part. When working with confidence intervals, you were limited to what you could calculate. It was basically a

Credible Intervals

95% confidence interval. Now, with Bayesian analysis, you can calculate any probability that makes sense for your article. For instance:

- There is a 95% probability that the BTP will receive between 14.15 and 20.8% of the vote.

Note the proper use of the word "probability" here. The credible interval *is* an interval about a probability.

But, we can calculate other probabilities, not just the credible interval. For instance, we also know:

- There is a 6.1% probability that the BTP will receive at least 20%.
- There is a 8.0% probability that the BTP will receive at most 15%.
- There is a 66.2% probability that the BTP will receive less than 18%.

... and any other probability that makes sense for the given election analysis. It is up to the researcher/author to determine what probabilities are interesting.[2]

Of course, I had to use the computer to perform these calculations. Here is the code used to obtain these results:

```
qbeta(c(0.025,0.975), 87,415)    ## 95% credible interval

1-pbeta(0.20, 87,415)            ## Pr[pi >= 0.20]
pbeta(0.15, 87,415)              ## Pr[pi <= 0.15]
pbeta(0.18, 87,415)              ## Pr[pi <  0.18]
```

Illustrative Bayesian Example 2

Next, to illustrate the importance of the analyst's prior information, let us use an informative prior distribution. For the sake of argument, let us pretend that I believe BTR will receive 20% of the vote, with a standard deviation of 1%. That is, I think there is a 95% chance that BTR will win between 18 and 22% of the vote.

With this expected value and standard deviation, I have R calculate a and b to be 319.8 and 1279.2, respectively. I got this from running `betaPrior(0.20,0.01)`.

---

[2] Given the structure and importance of the election, there is another probability that would be interesting: What is the probability that the BTP will make it to the second round? At this point, we cannot answer this question. All of the calculations so far have been about a single party. The important probability, however, requires comparing two. We will take up this probability in Chapter 5. After all, I have to save *something* for later, right?

Thus, the posterior distribution will be BETA(319.8 + 86, 1279.2 + 414) = BETA(405.8, 1693.2). And so, with this as our posterior distribution, we have that

- there is a 95% probability that the BTP will receive between 17.67 and 21.05% of the vote
- there is a 21.8% probability that the BTP will receive at least 20%
- there is almost a 0% probability that the BTP will receive at most 15%
- there is a 5.9% probability that the BTP will receive less than 18%

Here is the code used to obtain these results:

```
qbeta(c(0.025,0.975), 405.8,1693.2)   ## 95% credible interval
1-pbeta(0.20, 405.8,1693.2)           ## Pr[pi >= 0.20]
pbeta(0.15, 405.8,1693.2)             ## Pr[pi <= 0.15]
pbeta(0.18, 405.8,1693.2)             ## Pr[pi <  0.18]
```

Note: *This illustrates the importance of selecting an appropriate prior distribution. If the analyst has little prior knowledge of the population proportion, then a low-information prior (like the standard Uniform distribution) would be appropriate. However, if multiple polls strongly suggest that π should be in a small band, then a high-information prior would be appropriate.*[3]

From this discussion, one may think that a high-information prior may always be inappropriate. This is not necessarily the case. The prior should contain all information known by the analyst—and no more. When done properly, a high-information prior can help to mitigate the effects of a **poorly-performed poll**. Large differences between the prior expected value and the observed sample proportion may actually reflect a bad poll.

Unfortunately, it could also reflect a poorly chosen prior distribution.

Quick Example

Quinnipiac University held a poll of 807 Democrats and Democratically-leaning independents from August 1 until August 5 by contacting both landline and mobile phones. In that poll, 32% stated they preferred Vice President Biden over the rest of the Democratic field (258 for Biden

---

[3]Note that the frequentist results are equivalent when a "BETA(0,0)" is used as the prior distribution. Thus, even frequentists make assumptions about the prior distribution. It is just that Bayesians are explicit in their assumptions.

and 549 for others).[Qui19] This information allows us to calculate the following 95% intervals:

Interval	Lower Bound	Upper Bound
Confidence	0.285	0.355
Credible, uniform prior	0.288	0.353
Credible, BETA$(a = 25, b = 75)$ prior	0.282	0.343

Note: *The three estimates are quite similar. This is because the "high-information prior" is not too "high-information." It only contains an additional $25 + 75 = 100$ pieces of information about π. This is not large compared to the sample size of $n = 1016$. The low-information prior adds even less information, only $1 + 1 = 2$ pieces.*

You may want to return to this example after Chapter 5, when you have more experience with Bayesian analysis.

3.4 Conclusion

This chapter built on the previous by extending what a poll can tell us. Last chapter, we only calculated a point estimate (sample proportion). In this chapter, we looked at three ways to indicate the precision of that estimate: margin of error, confidence interval, and credible interval. Of these three, the first two are staunchly in the realm of frequentist statistics, while the third is a part of Bayesian analysis. Each of the three has strengths and weaknesses. The one you decide on will depend on your audience and what story you want to tell.

The margin of error explicitly states the uncertainty caused by the sampling scheme used. It puts that uncertainty at the center of the story. Confidence intervals focus on the parameter being estimated. They are sets of reasonable values for π. These allow the writer to focus on reasonable outcomes for the election, while still indicating uncertainty in the outcome.

The odd-one-out is the credible interval. It actually provides a statement of probability about the support level. Neither the margin of error nor the confidence interval can do this. More importantly, the theory underlying credible intervals allows one to both incorporate prior information *and* make statements about other interesting quantities.

While frequentist statistics is entrenched in introductory statistics courses, and thus in the mindset of the audience, there needs to be emphasis given to Bayesian analysis. There is so much that can be drawn from it.

3.5 Extensions

In this section, I provide some practice work on the concepts and skills of this chapter. Some of these have answers, while others do not.

Review

1. What is the relationship between the margin of error and the confidence interval?
2. What are the main difference between a confidence interval and a credible interval? When should you use each?
3. List several advantages and disadvantages to Bayesian analysis.
4. What is "coverage," and what is it used for?

Conceptual Extensions

1. In the beta distribution, one can think of $a+b$ as the effective sample size, n. With this, explain why $\mathbb{E}[\pi] = \frac{a}{a+b}$ makes sense.
2. Compare the variance of a beta distribution with that of the sample proportion, $\pi(1-\pi)/n$.

Computational Extensions

1. For the synthetic TKS Polling survey of Bhutan, calculate the probability that the People's Democratic Party (PDP) has

    (a) more than 40% support in the population
    (b) less than 50% support in the population
    (c) between 45% and 55% support in the population
    (d) has more than 30% support in the population

    For each of these, use the Beta prior with the following prior information

    (a) no prior information: $a = 1$, $b = 1$
    (b) relatively strong prior: $a = 150$, $b = 150$

(c) relatively strong prior: expected value of 50% and standard deviation of 3%

(d) relatively strong prior: expected value of 45% and standard deviation of 1%

3.6 Chapter Appendix

This appendix contains some "review" material on random variables and their mean (expected value) and variance.

3.6.1 R Functions

No statistician does these calculations by hand. We all use the computer. It helps to increase our precision and reduce the time needed to do the analysis. Here are some functions that help with the analyses covered in this chapter.

- betaPrior(a,b)

  This function allows one to easily calculate the a and b for the Beta prior distribution. The function's arguments are the expected value and standard deviation. The code for this function is provided in Section 5.8.1.1.

- isBetween(x, lower, upper)

  This function tests if the value x is between lower and upper. It returns TRUE if it is and FALSE otherwise. These can easily be cast to 0 and 1, allowing one to calculate the proportion true using the mean function.

- pbeta(x, a, b)

  This function calculates the cumulative probability (CDF) of x from a BETA(a,b) distribution. In others words, if $X \sim$ BETA(a,b), then this function calculates $\mathbb{P}[\,X \leq x\,]$. It can calculate the probability of support being at most a specific value x.

- 1 - pbeta(x, a, b)

  This is not a separate function, just a separate probability that may be of interest to you. This will calculate the complement of the CDF. That is, if $X \sim$ BETA(a,b), then this function calculates $\mathbb{P}[\,X > x\,]$. It can be used to calculate the probability of support being greater than a specific value x.

- qbeta(p, a, b)

  This function calculates the quantile of p from a BETA(a, b) distribution. In others words, if $X \sim \text{BETA}(a, b)$, then this function calculates x such that $\mathbb{P}[\, X \leq x \,] = p$.

3.6.2 Experiments

Experiments are at the heart of statistics. With experiments, one can gain a better understanding of the effects of randomness on what we are trying to model.

The Meaning of Coverage

So, what does it actually mean to be a 95% confidence interval? Not as much as one might think. Quite frankly, the best interpretation I can come up with is that it is a set of "reasonable" values for the population parameter. A given confidence interval does *not* have a 95% probability of containing π. At best, we can say that if we perform the experiment an infinite number of times, then 95% of those calculated confidence intervals will contain π. We do not know which ones, but we do know that 95% of them do. To help you have a better understanding of the meaning of a confidence interval, let us perform this experiment:

Experiment: Let us draw a random sample from a Binomial distribution, calculate the endpoints of the 95% confidence intervals for the sample proportion and for the Agresti-Coull estimators, and determine what proportion of those confidence intervals contain the population proportion (the coverage rate). We would want a coverage rate close to 95% because this is the meaning of a 95% confidence interval.

The code:

```
### Initialization
n  = 12
pi = 0.700

confLevel = 0.95
alpha = 1-confLevel
Za = abs(qnorm(alpha/2))

# Set aside internal memory
acMiss = spMiss = numeric()
lbAC = ubAC = numeric()
lbSP = ubSP = numeric()
```

Chapter Appendix

```
### Begin experiment
for(i in 1:1e6) {
  sample = rbinom(1, size=n, prob=pi)   ## The sample

# Agresti-Coull
  acEst = (sample+1)/(n+2)
  lbAC[i] = acEst - Za * sqrt(n*acEst*(1-acEst))/(n+2)
  ubAC[i] = acEst + Za * sqrt(n*acEst*(1-acEst))/(n+2)
  acMiss[i] = (pi<lbAC[i] || pi>ubAC[i])

# Sample proportion
  spEst = sample/n
  lbSP[i] = spEst - Za * sqrt(n*spEst*(1-spEst))/(n)
  ubSP[i] = spEst + Za * sqrt(n*spEst*(1-spEst))/(n)
  spMiss[i] = (pi<lbSP[i] || pi>ubSP[i])

}
# End experiment

### Results
1-mean(acMiss)   ## Coverage under Agresti-Coull
1-mean(spMiss)   ## Coverage under sample proportion
```

When I run this code, I get a coverage of 94.7% for the Agresti-Coull intervals, but only 87.6% for the sample proportion intervals. Thus, for a sample size of $n = 12$ and a population proportion of $\pi = 0.70$, the Agresti-Coull-based confidence intervals are superior to the sample proportion-based confidence intervals in terms of coverage.

Changing n and pi allows us to continue the experiment for different sample sizes and population proportions. I discovered that when π was near 0.500, there was little to no difference in the coverage between the two estimators. Also, I discovered that smaller sample sizes increases the differences between the two. Thus, since polls tend to deal with sample sizes in excess of $n = 400$, the sample proportion and the Agresti-Coull estimator perform similarly.

◇ ◇ ◇

The Main Lesson

Figure 3.3 shows something quite interesting. The Agresti-Coull intervals that do not contain $\pi = 0.70$ tend to be a subset of the sample proportion intervals that do not. In other words, the main lesson from this experiment is that using the Agresti-Coull estimator helps little once the number of people surveyed is large enough (greater than $n = 500$, perhaps). Beyond that point, there is no real benefit. Thus, returning to the sample proportion as our estimator is entirely appropriate.

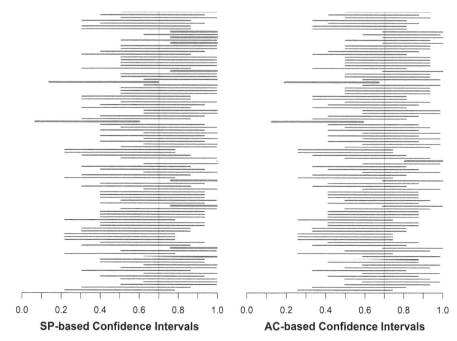

FIGURE 3.3
Illustration of the confidence intervals produced by both methods. The left panel is for the sample proportion-based confidence intervals; the right, Agresti-Coull. The thick confidence intervals do not contain the true $\pi = 0.70$; the thin intervals do. Both contain the first 100 confidence intervals calculated from a million iterations.

By the way, Figure 3.3 also illustrates why it is called "coverage." The segments represent the confidence intervals for each random draw from the population. The vertical dashed line is the true value of $\pi = 0.70$. When π is in the confidence interval, it is "covered" by it. This should happen 95% of the time if we are using a 95% confidence interval.

Comparing Coverage

This experiment was used in the chapter (Section 3.2.2) to learn more about the quality of the three methods for calculating the endpoints of a confidence interval.

Experiment: To compare the three methods for different polls, just change

Chapter Appendix

n and pi to reflect the poll and what you expect the population proportion π to be.

```
### Specify the parameters of the poll
n  = 500
pi = 0.20

Z = 1.96

### Generate the million samples (successes)
x = rbinom(1e6, size=n, prob=pi)


### Perform the calculations
p = x/n

# The First Estimator
lcl1 = p - Z * sqrt(0.50*(1-0.50)/n)
ucl1 = p + Z * sqrt(0.50*(1-0.50)/n)

## The Second Estimator (Wald)
lcl2 = p - Z * sqrt(p*(1-p)/n)
ucl2 = p + Z * sqrt(p*(1-p)/n)

### The Third Estimator
den  = 2 * (Z^2 + n)
num1 = 2*n*p + Z^2
num2 = sqrt( -4*n*p^2*Z^2 + 4*n*p*Z^2 + Z^4 )

lcl3 = (num1-num2)/den
ucl3 = (num1+num2)/den


### The Results: Coverage Rate
mean(isBetween(pi, lcl1,ucl1))
mean(isBetween(pi, lcl2,ucl2))
mean(isBetween(pi, lcl3,ucl3))
```

In running this for different values of π, you should see that the third confidence interval is always the best of the three, especially when π is not close to 0.50. Additionally, with different values of n, you will see that the improvement disappears as the sample size increases.

So, why is this third confidence interval (Eqn. 3.6) not taught in introductory statistics? Can you imagine trying to learn that as a first-year college student?

◇ ◇ ◇

4
Stratified Sampling

Polling houses conduct multiple polls throughout the election cycle. Each poll is designed to provide an estimate of the current support level for the candidate (or the position) or to predict who (or which position) will win the election. Previously, we examined one method for estimating support using a rather simple sampling method. Simple random sampling (SRS) produces estimates that are unbiased. However, as we saw, there is more to determining the poll's "goodness" than simply ensuring it is unbiased. Its mean squared error (combination of bias and variability) is a better measure of how well a particular poll performs.

In this chapter, we investigate ways of reducing the mean squared error. The main method is to use "stratified sampling" as the sampling scheme. While the estimator is biased under realistic operations, its mean squared error tends to be smaller than that of the SRS method, thus making it the superior estimator.

United States of America: The Presidential Election of 2016
In 2016, the United States held its 58th presidential election. Having already served two terms, Barack Obama was not able to stand for re-election. While this led to a large field of candidates vying for the presidency, the US party system had pared the candidates to one in each party. The two major-party candidates were Hillary Clinton (Democratic) and Donald Trump (Republican).

For much of the post-convention period, Clinton led Trump by a significant margin. However, as election day loomed (November 8, 2016), the race became tighter. In the final week, Monmouth University estimated the support for the two candidates.[Mon16] In that poll, Monmouth interviewed 802 registered voters. Half of those 802 were contacted using random-digit-dialing, with 201 from landline and 200 from cellular phone. The other half were taken from a list of registered voters (also 201 from landline and 200 from cellular phones).

TABLE 4.1
The weights used by Monmouth University in their final election poll (November 3–6).[Mon16] These are the population proportions (of the future 2016 voters) assumed by the analysts at Monmouth.

Age Range:		Race:	
18-34	25%	White	71%
35-49	25%	Black	14%
50-64	28%	Hispanic	11%
65+	22%	Asian/Other	5%
Partisanship:		**Gender:**	
Democrat	34%	Female	47%
Independent	38%	Male	53%
Republican	28%		

The results were weighted according to age range, race, partisanship, and gender with population proportions estimated from the voter list and the US Census Bureau. Table 4.1 provides the weights (assumed proportions in the population) used by Monmouth University. This poll resulted in a margin of error of 3.6 percentage points for the entire population, but larger margins of error for subsets of the population (for each gender, the margin of error is approximately 5.0 percentage points).

From this poll, Monmouth University estimated Clinton support between 45.5% and 52.5% and Trump support between 38.5% and 45.5%. In the election a few days later, of the 136,669,276 total votes cast, 65,853,514 were for Clinton (48.2%) and 62,984,828 were for Trump (46.1%).[New17] While Clinton's actual popular vote share was within the confidence interval, Trump's was not. He over-performed according to the poll.

Regardless of Clinton receiving more votes, Trump won the election. This is because the United States uses an Electoral College to elect the president. The members of the Electoral College are, in turn, elected by voters at the state level. The number of Electors for each state is equal to its number of Representatives and Senators.

For instance, the number of Electors from Wyoming is 3, while the number from Illinois is 19. This election method was created because the United States is a federation in which the central government and the states all have fundamental political power. Thus, the number of Electors from a state is based on both its population (Representatives) and its existence as a state (Senators).

As such, even though Clinton received more of the popular vote, Trump received 304 of the 538 electoral votes cast. Thus, Trump won the election 2016. Because of the Electoral College, it is important to focus on the vote at

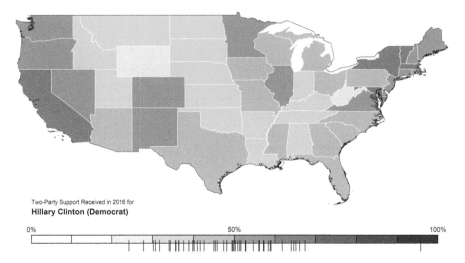

FIGURE 4.1
Final results of the 2016 US presidential elections across the 51 voting polities (50 states, plus the District of Columbia). The darker the shade, the higher the proportion of votes for Clinton, the Democratic candidate. These data are from Wikipedia.[Wi24a]

the state level when modeling the US Presidential elections. The proportion of the votes for Clinton at the state level are illustrated in Figure 4.1.

Overview

Previously, we covered simple random sampling. In that sampling scheme, each person in the population has an equal chance of being selected. In using this scheme, we showed that the natural estimator is unbiased and has variance $\mathbb{V}[\,P\,] = \pi(1-\pi)/n$, where P is the sample proportion, π is the population proportion, and n is the size of the sample.

We also showed that the Agresti-Coull estimator was *biased* but usually had a lower mean squared error (MSE) than the sample proportion for smaller sample sizes and middling values of π; that is, the Agresti-Coull estimator was more **concentrated** around π than was the sample proportion, thus illustrating the importance of the MSE over the bias as a measure of estimator quality.

In this chapter, we introduce a sampling scheme designed to produce an estimator with a lower mean squared error than the sample proportion under simple random sampling (SRS). This new estimator/sampling scheme does

4.1 Stratified Sampling

Let us suppose that I would like to estimate the proportion of Americans who agree with the statement:

> "Women should receive an entire year of paid maternity leave."

To estimate this proportion, I contact 1000 people using random digit dialing.[1] In that sample, 537 stated that they supported the position. With this, I send a press release to a prominent international daily that states, in part,

> Ole J. Forsberg, a researcher at Knox College in Illinois, polled 1000 Americans and estimates that 53.7% of the US population—a clear majority—believes that women should receive a year of paid maternity leave.

However, one has to wonder about the structure of my sample. It is clear that there may be a relationship between gender and position on this particular issue. If my sample has a higher proportion of females than are in the population, then my estimate will naturally be biased toward the "female" position; it would over-sample females. Similarly, if my sample has too many males, then it will be biased toward the "male" position.

> **Aside.** *To be clear, stating that there is a relationship between two variables does not imply that* all *women agree with "Women should receive a year of paid maternity leave" nor that* all *men disagree with it. It merely means that the population proportion of females in favor of the position is not the same as that of males.*

This issue becomes more evident when I share that my particular sample had 300 males and 700 females. Of those 300 males, only 117 agreed with

---

[1] In "random digit dialing," a telephone number is randomly generated by a computer. If it is necessary to focus on the survey on a specific exchange, then only the final four digits are randomized. If a certain US state is the focus, area codes can be randomized. Similarly, if the survey is international, then country codes (and below) are randomized.

the position (39%), while 420 of the 700 females did (60%). This is a rather large "gender gap" in this sample—significant evidence of a gender gap in the population.[2] It is also significant evidence that the researcher needs to take gender into consideration in estimating the proportion of Americans in favor of the statement. So, let's do that.

According to World Bank data,[Wor19] the proportion of females in the United States is 50.5%. Since my poll sampled much more than 50.5% female, it is clear that my particular poll is biased toward the "female" point of view. If we simply report the proportion in the poll who agree with the statement, then it would be a biased estimate—biased toward the "female" position.

The simplest method to mitigate this bias is to weight the results based on "female-ness." In general, weighting is an important technique in sampling, as it helps to make the sample much more representative of the target population. In essence, it makes the over-represented population less important—and the under-represented population more important—in the sample.

Thus, since we "know" that the proportion of females in the population is 50.5%,[Wor19] we weight our female sample by 0.505 and our male sample by 0.495. This provides an adjusted estimate of

$$\text{Estimate} = (0.505)(0.60) + (0.495)(0.39) = 0.49605$$

Thus, a press release taking into consideration the sample's structure could be

> Ole J. Forsberg, a researcher at Knox College in Illinois, polled 1000 Americans and weighted according to gender, estimates that 49.6% of the US population—a minority—believes that women should receive a year of paid maternity leave.

To foreshadow a problem with stratified sampling, the US Census Bureau estimates that the proportion of females in the population is 50.8%.[US 19] Thus, using the US Census estimate, the press release could be

> Ole J. Forsberg, a researcher at Knox College in Illinois, polled 1000 Americans and weighted according to gender, estimated that 49.7% of the US population—a minority—believes that women should receive a year of paid maternity leave.

---

[2] As with most things in statistics, there is a procedure that can be used to test if there is enough evidence that this gap exists in the population. The two-sample proportions-procedure works quite well for large sample sizes. In practice, sample sizes in excess of 200 may be needed. The appendix (page 92) has an experiment that explores this. As always, experimenting helps understand the effects of our assumptions.

However, since the proportion of the 2016 voters who were female was 53%, the press release could also have stated

> Ole J. Forsberg, a researcher at Knox College in Illinois, polled 1000 Americans and weighted according to gender, estimates that 50.1% of the voting US population—a majority—believes that women should receive a year of paid maternity leave.

Also, if I estimate that the 2024 electorate will be weighted 60% female, the press release could also have stated

> Ole J. Forsberg, a researcher at Knox College in Illinois, polled 1000 Americans and weighted according to gender, estimates that 52.6% of the 2024 voters—a clear majority—believes that women should receive a year of paid maternity leave.

All conclusions are supported by the data. The differences in the conclusions are a result of the particular population chosen and how the particular analyst/researcher/polling house estimates the weights.

> **Aside.** *This exercise is not just academic. The effect of not knowing important characteristics of the population can be quite severe in practice. In 2016, the New York Times had Siena College poll Floridians about their choice for president.[Coh16] The data consisted of the preferences and demographics of 867 citizens. The New York Times sent the raw data to four different polling firms to investigate the "house effect" on the estimate.*
>
> *The estimates ranged from Clinton +4% to Trump +1%. The wide variation was mostly due to how the polling houses weighted the demographic groups. The article makes* very *interesting reading.*

The variation in estimates was *also* due to how they determined who was going to vote in the election (the "likely voter"). Remember that election polls tend to estimate which candidate will be elected. Thus, the population of interest is the "voter," which does not yet exist for the future election.

4.2 The Mathematics of Estimating π

Now that we have explored stratified sampling a bit, let us gain some additional understanding of it using mathematics. The stratified sampling estimator, $\overline{X}_{str}$, is defined as

$$\overline{X}_{str} := \sum_{g=1}^{G} w_g \overline{X}_g$$

Here, G is the number of groups (strata) being used, $\overline{X}_g$ is the mean of the sample drawn from Stratum g, and w_g is the proportion of the *population* belonging to Stratum g as claimed by the researcher.

Specifically, in the previous example, $G = 2$, $\overline{X}_1 = 0.60$, and $\overline{X}_2 = 0.39$. The various female weights, w_1, were either $0.505, 0.508, 0.530, 0.600$, or 0.700 depending on the weighting used, and the male weights were $w_2 = 0.495, 0.492, 0.470, 0.400$, or 0.300 depending on the female weighting used.

Reading Note

This section becomes rather mathy rather quickly. While I hope that you skim through this to see the final results regarding bias, my feelings will not be *too* hurt if you skip over the math and focus on the conclusions about the appropriateness of stratified sampling.

4.2.1 Bias

Recall from Section 2.2.1 that an estimator's bias is defined as the average divergence from the population value (Defn. 2.1); that is, the bias of a generic estimator P of π is $\text{bias}[\, P \,] := \mathbb{E}[\, P - \pi \,] = \mathbb{E}[\, P \,] - \pi$. We use this definition to calculate the bias of the stratified sampling estimator:[3]

$$\text{bias}\!\left[\, \overline{X}_{str} \,\right] = \mathbb{E}\!\left[\, \overline{X}_{str} \,\right] - \pi$$

---

[3] In this proof, N_g is the number of people in the *target population* belonging to Stratum g, X_g is the number of people in the *sample* in Stratum g who support the position, N is the population size, and π_g is the proportion of the population in Stratum g who support the position.

Normally, proofs go in the appendix. However, I think I need to "show the math" to emphasize the importance of knowing the population. If the proportion of males and females in the population are now known, for instance, then this method is biased. How much? Keep reading.

$$\text{bias}\left[\overline{X}_{str}\right] = \mathbb{E}\left[\sum_{g=1}^{G} w_g \overline{X}_g\right] - \pi$$

$$= \left(\sum_{g=1}^{G} w_g\, \mathbb{E}[\,\overline{X}_g\,]\right) - \pi$$

$$= \left(\sum_{g=1}^{G} w_g\, \pi_g\right) - \pi \qquad (4.1)$$

$$= \left(\sum_{g=1}^{G} \frac{N_g}{N}\frac{X_g}{N_g}\right) - \pi \qquad (4.2)$$

$$= \left(\sum_{g=1}^{G} \frac{X_g}{N}\right) - \pi$$

$$= \left(\frac{X}{N}\right) - \pi$$

$$= \pi - \pi$$

$$= 0$$

Thus, the stratified estimator is unbiased... as long as these steps are all mathematically correct.

The only place where an error may enter is from Eqn. 4.1 to Eqn. 4.2. If it *is* true that $w_g = N_g/N$, i.e. that the weights applied by the researcher is the same as the proportion of that stratum in the population, then the steps are all mathematically true. If this equality is *not* true, i.e. if the weights are wrong, then the stratified estimator is **biased**.

In Other Words

The stratified estimator is unbiased *only when* the weights used, w_g, are correct, when w_g is actually the proportion of the population belonging to Stratum g. If the weights are *not* correct, then the stratified estimator is biased.

Thus, referring to the Monmouth Poll discussed on page 69, if the proportion of the voters who are Democratic is not 34%, then its estimate is biased. Since the actual proportion of voters in the 2016 election who were Democratic was actually 37%, the estimates are biased. They were biased against Democratic voters.

Unfortunately, this requirement is rarely (if ever) perfectly met in practice. For instance, in the previous section, we had five good estimators of the population proportion of females: 0.505, 0.508, 0.530, 0.600, and 0.700. Which is correct? The one you use affects your estimates. Usually the effect on the final estimate is small; however, if the stratified estimate is close to 50%, the one you use may determine your conclusion as to which position is the majority position—or which person will win the election.

Even more problematic, polling houses tend to stratify on more than just one variable. They may use such variables as gender, political party affiliation, and socio-economic status (SES) to better estimate the support level for the position. In such a case, this means going beyond simply knowing the proportion of the population that is female. One must know the proportion of the population that belongs to the group female-high SES-Republican; to the group male-medium SES-Democrat; to the group female-low SES-Republican; as well as all other possible combinations.

Furthermore, if the population of interest is the voters in an *upcoming* election, the polling houses must estimate the weights in each combination of variable categories for a population **that does not yet exist**. This means that polling firms need to first create a method for determining *future* voters, then stratify according to the variables in that particular population.

Do you see the difficulty with estimating vote results? It is much easier to estimate positions about the entire nation than it is to estimate things about those who *will* actually vote. In fact, it is the methods for creating these estimates that leads to certain polling houses having a reputation of being "left-leaning" or "right-leaning."

The Fundamental Problem

For estimating election outcomes, the fundamental problem is that there is no way of knowing the *correct weights* before the election takes place *and* there is no currently agreed-upon way of incorporating that uncertainty into confidence intervals (Chapter 3).

Bayesian methods (Chapter 5) offer advantages over frequentist methods in this case. They allow one to specify a probability distribution on the various weights. The resulting credible intervals are "easily" calculated from that. Such methods are not widely used, however.

For instance, using Bayesian analysis, we could specify priors for each of the demographic groups like

$$P_f \sim \text{BETA}(a = 330, b = 295)$$

for the proportion of the voting population that is female. This distribution leads to an expected vote that is female of $53 \pm 2\%$; that is, we are 95% sure

The Mathematics of Estimating π

that the proportion of the voters who are female is between 49% and 57%.[4] Looking at the historical trends, this seems reasonable for the 2024 election in the United States.

"Reasonable," not "True"

Note that I stated it was "reasonable," not that it was true. We will not know "true" the actual results until the votes are tallied. Such is the nature of a random variable and the goal of statistics.

4.2.2 Investigating Bias

Section 4.2.1 shows that the stratified sampling estimator is biased unless the correct weights are used. However, the important question really is this:

"Is the amount of bias really that important?"

The question of *practical* significance is always an important question to explore. It is fine to speak of bias, but it is more important to be able to provide reasonable bounds on that bias. The following experiment investigates the bias of the stratified estimator.

Experiment: This experiment looks at the effect of those population proportion estimates on the bias of the stratified estimator. In this experiment, we will look at the estimate when we vary the estimated value of the proportion of females in the population, w_f, from 0% to 100%.

```
# Survey results
pf = 0.60     ## Proportion of females in favor
pm = 0.39     ## Proportion of males in favor

# Calculate stratified sampling estimates
wf = seq(0, 1, 0.01)       ## female weights from 0 to 100%
est = pf*wf + pm*(1-wf)    ## the stratified estimates

# Plot of estimate as a function of the weight
plot(propFemale, est)
```

---

[4]This is as good a place as any to state this. When there is a value following the estimate, such as 53±2%, it is not necessarily clear what that ±2% means. The 2% could indicate the standard deviation or the margin of error at 95%. Hopefully, the source will be sufficiently clear which. In *this* case, I am using it as a standard deviation. This allows the reader to "easily" determine the margin of error according to their own confidence level.

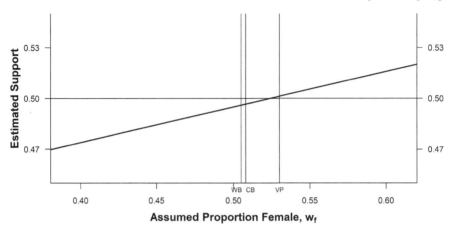

FIGURE 4.2
A graphic illustrating the effect of the assumed proportion of females in the population on the estimated support for "Women should be treated the same as men in the military." Note that "majority support for this position" depends on the estimated proportion of females used. The "WB," "CB," and "VP" refer to the World Bank, Census Bureau, and 2016 voting population of females in the population of interest.

The logic of this code is to generate many possible values for w_f, calculate the stratified sampling estimate based on those weights (and the results of the survey, $p_f = 0.60$ and $p_m = 0.39$), then plot those results. These results are graphed as Figure 4.2.

Note that the range of estimates is unsurprisingly between 39% and 60%. The former happens when we assign a weight of $w_f = 0$ (claim there are no females in the population); the latter, when we assign $w_f = 1$ (claim there are no males in the population). Clearly the extremes make no sense for a general population. Two sources estimate w_f to be 0.505 and 0.508 in the general population.[US 19, Wor19] These produce estimates of 0.496 and 0.497, respectively.

◇ ◇ ◇

However, what is the support amongst those who *will* vote in the next election? Since that event has yet to happen, it is not clear what value should be used for w_f. Should we use $w_f = 0.530$ to reflect the gender-ratio of the 2016 election? Do we think females will turn out to vote at a lower ratio than before? If so, by how much? Since all of the estimated support levels are close to 50%, the one we select will determine our ultimate conclusion.

The Mathematics of Estimating π 79

In other words, one use for this graphic is to emphasize that the outcome of the election depends on the turnout of women *as compared to men*. According to Figure 4.2, if females comprise more than 52.4% of the voters in the election, then their position becomes the majority position. Anything less than that and it is the minority position.

Polling Houses Differ

Again, how the polling house answers these questions about who will vote determines, in large way, the final estimates made by the house. This issue is compounded by the fact that polling houses stratify on more than just gender. Differences in these estimates cause differences in the final vote estimate made by the polling house.

4.2.3 The Goodness of the Stratified Estimator

Recall that the bias tells a small part of the story. Bias only speaks to the large-scale behavior of an estimator; the mean squared error (MSE) is a better measure of how likely it is for a specific estimate to be close to the true value. This is because the MSE is a measure of how concentrated the estimator is around the parameter's true value (Section 2.2.3). Recall that larger values of the MSE indicate the average distance between the estimate and the true value is large; smaller values of the MSE indicate the average distance between the estimate and the true value is small, that the estimates tend to be close to reality. This is the strength of the stratified estimator; its mean square error tends to be smaller than that of the SRS estimator (Section 2.2). In this section, let us derive the MSE for the stratified estimator and explore when it is better then the SRS estimator.

Reading Note

The upshot of the math in this section is that we are able to derive a formula for the mean squared error of the stratified estimator. From that, we are able to see *when* the stratified estimator is better than the simple random sampling estimator. Specifically, it tells us how close the weights have to be to reality.

The MSE of the Estimate

The mean squared error is the sum of the estimator's variance and the square of its bias. Here, we calculate the variance of the estimator.

$$\mathbb{V}[\,\overline{X}_{str}\,] = \mathbb{V}\left[\sum_{g=1}^{G} w_g \overline{X}_g\right]$$

$$= \sum_{g=1}^{G} \mathbb{V}[\,w_g \overline{X}_g\,]$$

$$= \sum_{g=1}^{G} w_g^2\, \mathbb{V}[\,\overline{X}_g\,]$$

$$- \sum_{g=1}^{G} w_g^2 \frac{\pi_g(1-\pi_g)}{n_g}$$

With this, we know that the MSE of the stratified estimator is

$$\mathrm{MSE}\left[\,\overline{X}_{str}\,\right] = \left(\left(\sum_{g=1}^{G} w_g \pi_g\right) - \pi\right)^2 + \sum_{g=1}^{G} w_g^2 \frac{\pi_g(1-\pi_g)}{n_g}$$

By itself, it is rather difficult to compare this MSE to that of the SRS estimator

$$\mathrm{MSE}\left[\,\overline{X}_{srs}\,\right] = \frac{\pi(1-\pi)}{n}$$

In fact, the formula really makes it seem as though the SRS estimator is superior (see Section 2.2.3). It *certainly* is easier to work with.

The next experiment explores the differences in the mean square error between the two estimators.

Experiments in Statistics

One way of seeing the effect of being wrong on the weights is to do some rather heavy mathematics. Another way is to perform a few statistical experiments. Because these experiments provide skills that can be applied to a wide variety of cases (and they are low-math), this section walks through some experiments to explore the effect of being wrong on the weights.

The key for these experiments is to see how the assumptions are exhibited in the code and to see how the results are interpreted.

The Mathematics of Estimating π

To illustrate the difference in the MSE of the two estimators, we need to specify several values for our experiment. The population values, π and π_g, must be assumed. The other values are part of the survey design. The first experiment mimics simple random sampling; the remaining, various stratified sampling cases.

In all cases, the sample size is $n = 1000$ people with the population support $\pi = 0.500$. It is this population support that these methods are trying to estimate. To estimate the mean squared error (MSE), one can just generate a large number of "poll results," calculate the estimate for each poll, then find the average (mean) of the squared errors: `mean ( ( est - p ) ^ 2 )`.

This is done for each of the following situations.

Simple Random Sampling

Simple random sampling assumes the sample is taken from the entire population, where the probability of being selected is the same for each member of the population. Recall from Section 2.1 that the distribution of the number of successes in the poll is Binomial and that the usual estimator is the number of successes divided by the number of trials, the sample proportion: `x / n`.

```
p = 0.50; n = 1000                            ## Parameters

x = rbinom(1000000, size=n, prob=p)           ## Successes in sample
estSRS = x/n                                  ## SRS estimator

mean( (estSRS - p)^2 )                        ## Mean squared error
```

The final line provides the estimated mean squared error for this sampling scheme. According to this, the MSE is approximately 0.000 250. Since this is a simple sampling scheme, we can directly calculate the MSE to be $\pi(1-\pi)/n = 0.500(1 - 0.500)/1000 = 0.000\,25$ (see Section 2.2.3).

Phew! It's nice when the computer gives the right answer.

Stratified Sampling, I

Next, let us define the parameters for our stratified sampling poll. For the sake of consistency, it also consists of $n = 1000$ people but with them evenly divided into four groups: $G = 4$, $n_g = 250$. Let us specify our weights to all be $w_g = 0.25$.

Let these weights be *correct*; that is, the population actually consists of these four groups in equal amounts.

The following code translates those conditions.

```
### Population information
pi = c(0.50,0.50,0.50,0.50)      ## Real subpopulation support
pg = c(0.25,0.25,0.25,0.25)      ## Real weights (population)
p  = sum( pi*pg )                ## Total population support

### Sample information
ng = c(250,250,250,250)          ## Subsample sizes
wg = c(0.25,0.25,0.25,0.25)      ## Claimed weights

### Generate a million polls
x1 = rbinom(1e6, size=ng[1], prob=pi[1])   ## Sub-sample 1
x2 = rbinom(1e6, size=ng[2], prob=pi[2])   ## Sub-sample 2
x3 = rbinom(1e6, size=ng[3], prob=pi[3])   ## Sub-sample 3
x4 = rbinom(1e6, size=ng[4], prob=pi[4])   ## Sub-sample 4

### Calculate the stratified estimator for each poll
estSTR = wg[1]*x1/ng[1] + wg[2]*x2/ng[2] + wg[3]*x3/ng[3] +
    wg[4]*x4/ng[4]

### Mean squared error
mean( (estSTR-p)^2 )
```

According to this, the MSE is approximately 0.000 250. This is the same as the simple random sample estimator. This result illustrates this conclusion: If the groups support the candidate equally in the population (π_i is constant), then there is no advantage to using stratified sampling over simple random sampling. Actually, there is no disadvantage, either (other than the math).

Specifically, note that there is no relationship between the grouping variable G and the level of support, π_g. This illustrates the importance of selecting a grouping variable that *is* related to what you are trying to estimate. In the "parental leave" examples, the grouping variable is gender, and gender appears to be related to the level of support for the statement "Women should receive a year of paid maternity leave."

Stratified Sampling, II

The previous example assumed the candidate support was the same in the four sub-populations. Let's change this to showcase stratified sampling:

- Let the candidate support (π_i, `pi`) differ in the four groups.
- Let the four groups be equally represented in the population.
- Finally, let us still assume we chose our weights correctly (i.e., that `wg` and `pg` are equal).

The following code uses these conditions to estimate the MSE:

The Mathematics of Estimating π

```
### Population information
pi = c(0.95,0.35,0.25,0.45)    ## Real sub-population support
   <-
pg = c(0.25,0.25,0.25,0.25)    ## Real weights (population)
p  = sum( pi*pg )              ## Total population support

### Sample information
ng = c(250,250,250,250)        ## Chosen sub-sample sizes
wg = c(0.25,0.25,0.25,0.25)    ## Stated (claimed) weights

### Generate a million polls
x1 = rbinom(1e6, size=ng[1], prob=pi[1])   ## Sub-sample 1
x2 = rbinom(1e6, size=ng[2], prob=pi[2])   ## Sub-sample 2
x3 = rbinom(1e6, size=ng[3], prob=pi[3])   ## Sub-sample 3
x4 = rbinom(1e6, size=ng[4], prob=pi[4])   ## Sub-sample 4

### Stratified estimator:
estStS = wg[1]*x1/ng[1] + wg[2]*x2/ng[2] + wg[3]*x3/ng[3] +
         wg[4]*x4/ng[4]

### Mean squared error
mean( (estStS-p)^2 )
```

Note that the support in the sub-populations differ. Sub-population 1 has support $\pi_1 = 0.95$, while $\pi_2 = 0.35$, $\pi_3 = 0.25$, and $\pi_4 = 0.45$. (Note that the total population support remains 50%. This aids in comparing results across experiments.)

With this, the estimated mean squared error is 0.000 177, which is about 30% less than the SRS mean squared error. And so, when there *is a relationship* between the group and the support level, the stratified estimator has a lower MSE and is the better estimator.

Note, however, that this experiment still assumes that the analyst is correct about the weights applied, that $w_g = N_g/N$. We already know that the stratified estimator is biased if this equality is not true. The next two experiments look to see how close the weights have to be to the population proportions to be "close enough" in the resulting estimates

At Least Two More Stratified Sampling Experiments

The third and fourth experiments examine the MSE when the researcher is wrong on the assigned weights, $w_g \neq N_g/N$. Because the code is quite similar to the above code snippets, I put them in the appendix to this chapter (starting on page 96).

The third experiment explores what happens when the stated weights do *not* match reality. Note that in the previous two experiments, the stated weights match the population (true) weights. This experiment explores what happens when the weights are severely wrong.

According to this experiment, the MSE is approximately 0.014 53. This is *much greater* than the MSE of the SRS estimator—more than 50 times higher. Thus, if the weights are this far from reality, the stratified estimator is much worse than the SRS estimator.

Note that the claimed weights differ from reality by as much as 20 percentage points. It is very unlikely that a good analyst will miss the population proportions by that much. It is more likely that the differences will tend to be less than 5 points different. The next experiment looks to see what happens when they are off by no more than 5 points.

How bad is the MSE if the weights are all off by 5 percentage points in each stratum? The fourth experiment explores this condition (page 96). Being off by this much makes the MSE only *double* that of simple random sampling. The closer the weights are to reality, the lower the MSE of the stratified estimator.

I leave it as an exercise to see how far off one can be to make the MSE values roughly comparable. Modify the code of the fourth experiment. Keep changing the claimed weights (the wg line) until you get MSE values comparable to those of simple random sampling (MSE = 0.000 25). When you do this, you will find many situations in which the MSE for stratified sampling is comparable to that of simple random sampling.

When I did this, I discovered that being off by at most 2pp makes the MSEs similar. The MSE for the stratified estimator is 0.000 23, a bit less than that of the SRS estimator.

The lesson from these experiments is that the stratified estimator provides a lower mean squared error as long as the claimed weights are within about 2pp of the real weights. In other words, if your weights are close, then the stratified estimator does a better job at estimating the population proportion than does simple random sampling. This is the real reason polling firms use stratified sampling for their estimates. They only need to be close on their strata estimates to have a superior estimator.

Is 2pp Difficult?

How difficult is it for an analyst to be off on the weights by at most 2 percentage points? Under typical circumstances, it is not too difficult. Analysts know the population; it is their job. During election cycles, they pay close attention to the demographics. They also use surveys to inform their weights.

With that said, even the best analysts can be off by more than 2pp. The Monmouth University polling example below shows that even the best can miss, especially in elections where it is difficult to determine the voting population.

> Also, a lot of the error in the Brexit polling came down to misweighting for the actual population that voted (see Chapter 10). The rest of the error in this referendum arose from *not* stratifying along an important variable—educational attainment.

4.3 Confidence Intervals

Now that we have looked at calculating the point estimate, it is time to see how to calculate the confidence intervals. Note that there is an entire chapter dedicated to confidence intervals (and p-values). Chapter 3 covers calculations, interpretations, and meanings.

Thus, for the sake of interpretability, the confidence intervals using the stratified estimator follows the usual structure with the expected substitutions.

$$\text{endpoints} = \text{point estimate} \pm Z_{\alpha/2} \times \text{standard error}$$

$$= \left(\sum_{g=1}^{G} w_g \overline{X}_g \right) \pm Z_{\alpha/2} \times \sqrt{ \sum_{g=1}^{G} w_g^2 \frac{\hat{\pi}_g(1-\hat{\pi}_g)}{n_g} }$$

As discussed in Section 3.1.2, $\hat{\pi}_g$ can take on a few different values. To obtain conservative confidence intervals, use $\hat{\pi}_g = 0.500$. To obtain intervals that are usually closer to "correct," use $\hat{\pi}_g = p_g$, the sub-sample proportions. Neither is "correct," because correct requires actually knowing the value of π_g. However, if we knew those values, we would not have to use statistics to estimate π.

Example. Let us return to estimating the proportion of the US population that agrees with the statement "Women should receive a year of paid maternity leave" by stratifying on gender. From our poll, 60% of the females and 39% of the males agreed with the statement. Using the US Census estimates, the proportion of the US population that identifies as female is 0.505.

As calculated earlier, the point estimate is

$$\overline{X}_s = \sum_g w_g \overline{X}_g$$
$$= w_f \overline{X}_f + w_m \overline{X}_m$$
$$= (0.505)(0.60) + (0.495)(0.39)$$

$$= 0.49605$$

Thus, the endpoints of a 95% confidence interval are

$$\text{endpoints} = \text{point estimate} \pm E$$
$$= 0.49605 \pm E$$

Here,

$$E = Z_{\alpha/2} \sqrt{\sum_g w_g^2 \frac{\hat{\pi}_g(1-\hat{\pi}_g)}{n_g}}$$

$$= (1.96)\sqrt{w_f^2 \frac{0.5(1-0.5)}{n_f} + w_m^2 \frac{0.5(1-0.5)}{n_m}}$$

$$= (1.96)\sqrt{(0.505)^2 \frac{0.5(1-0.5)}{700} + (0.495)^2 \frac{0.5(1-0.5)}{300}}$$

$$= (1.96)\sqrt{(0.505)^2\, 0.000357 + (0.495)^2\, 0.000833}$$

$$= (1.96)\sqrt{0.0000911 + 0.0002042}$$

$$= (1.96)(0.01718)$$

$$= 0.033679$$

Thus, we are 95% confident that the support for the statement "Women should receive a year of paid maternity leave" is from 46.23% to 52.97%. This can also be written as 49.61 ± 3.37%.

Note that the confidence interval contains the value 50%. Thus, there is no significant evidence that a majority of the US population supports the "pro" position (the confidence interval contains values less than 50%). Similarly, there is no significant evidence that a majority of the US population supports the "con" position (the confidence interval also contains values greater than 50%).

Aside. *If we were to perform this experiment a gazillion times, 95% of the confidence intervals would contain the actual value of π (see Figure 3.3*

Confidence Intervals

on page 66). Unfortunately, we are not able to determine if this particular *confidence interval contains* π. Such is the problem using confidence intervals (see Chapters 3 and 5).

Note that if we were to ignore gender in the analysis and treat the data as if it were gathered using simple random sampling, we would be 95% confident that the population support for the statement "Women should receive a year of paid maternity leave" is from 50.60% to 56.80%, in other words 53.7±3.10%. The following code shows this.

```
x =   537      ## Successes
n =  1000      ## Sample size
phat = x/n     ## Point estimate

## Margin of error
E = -qnorm(0.025) * sqrt( 0.500*(1-0.500)/n )

## Confidence Interval endpoints
phat-E
phat+E
```

Note that this simple random sampling estimate is equivalent to a stratified estimator using $w_f = 0.700$.

Thus, the simple random sample estimator from Chapter 2 is clearly problematic. In the vast majority of cases, the proportion of females in the population of interest is not even close to 70%.

Example (Monmouth University). For our next example, let us return to the Monmouth University example that started this chapter. They report that, of the 802 registered voters in the sample, the number of females was 410; and males, 392. Also, the number of Democrats, Independents, and Republicans were 267, 298, and 231, respectively.[5] Table 4.2 summarizes this information, along with the support for Clinton in each sub-sample.[Wi20d]

The final two columns illustrate the differences between the assumed voting proportions (sample weight) and the actual population voting. While Monmouth did a fine job of estimating the gender ratios, it missed on the party ratios. It missed Democratic voting by 3pp (34% vs. 37%), Independent voting by 7pp (38% vs. 31%), and Republican voting by 5pp (28% vs. 33%). Note that Monmouth University's polling institute is first class. That it missed illustrates how hard it is to estimate a future population, especially a future *voting* population in a contentious election.

---

[5]This information is rarely made public in polls. Monmouth University, provides this information in the form of a linked "cross-tab."[Mon16] Thanks Monmouth! You are still awesome!!

TABLE 4.2
Cross tabulation summary of the Monmouth University report, assuming the two variables (gender and party affiliation) are independent. The second column is the support for Clinton in that sub-sample.

Sub-Population	Sample Support	Sample Proportion	Sample Weight	Population Proportion
Female	0.55	0.511	0.53	0.53
Male	0.43	0.489	0.47	0.47
Democrat	0.92	0.333	0.34	0.37
Independent	0.42	0.372	0.38	0.31
Republican	0.09	0.288	0.28	0.33

Using the gender information from Table 4.2, the estimated Clinton support nationwide is

$$0.53 \times 0.55 + 0.47 \times 0.43$$
$$= 49.36\%$$

with margin of error

$$1.96 \times \sqrt{0.53^2 \frac{0.5(1-0.5)}{410} + 0.47^2 \frac{0.5(1-0.5)}{392}}$$
$$= \pm 3.5$$

Using the party information, the estimated Clinton support is

$$0.34 \times 0.92 + 0.38 \times 0.42 + 0.28 \times 0.09$$
$$= 49.76\%$$

with margin of error

$$1.96 \times \sqrt{0.34^2 \frac{0.5(1-0.5)}{267} + 0.38^2 \frac{0.5(1-0.5)}{298} + 0.28^2 \frac{0.5(1-0.5)}{231}}$$
$$= \pm 3.5$$

Thus, by stratifying only on gender, the estimated support for Clinton in this poll is from 45.9% to 52.9%. Stratifying only on party affiliation, the interval is quite similar: from 46.3% to 53.3%. The reality is that she received 48.2% of the votes cast. Both intervals contain this value.[6]

---

[6] By the way, there is an R function that performs these calculations for you. Please check out Section 4.6.3 on page 94 for more details.

Is it possible to stratify on both gender and party information? Absolutely. It just requires us to know three things. First and second, the polling firm must keep track of the numbers in the sample that fit in both categories as well as their positions on the polled issue. This is not a problem as this information is recorded in the raw and can be obtained through basic tabling methods.

Third, and definitely problematic, we have to know those proportions in the population. This is not as easy to know as it sounds. One is tempted to just multiply the relevant proportions to obtain these. In other words, one may want to use the data in Table 4.2 to estimate the proportion of female Democrats in the population as $0.53 \times 0.37 = 0.1961$, the proportion of male independents as $0.47 \times 0.31 = 0.1457$, etc. This does seem eminently reasonable.

However, this makes the assumption that party affiliation and gender are *independent*, that the proportion of females who are Democratic is the same as the proportion of females who are Republican. In general, this is an untenable position. It is especially problematic in the United States because the "**gender gap**" in party affiliation is well documented:

> In addition to the gender gap in voter turnout, partisan preferences differ widely by gender. Pew Research Center survey data going back more than two decades shows a growing gender gap in partisan affiliation. In 2018 and 2019, the Democratic Party held a wide advantage with women: 56% of female registered voters identified as Democrats or leaned toward the Democratic Party, while 38% identified as Republicans or leaned toward the GOP.[Igi20]

4.4 Conclusion

This chapter concluded our introduction to polling methods. Chapter 2 had us assume we were collecting our data using simple random sampling. This method requires each member of the target population has the same probability of being selected. This either requires an infinite population or allowing for an individual to be selected multiple times. The former does not exist and the latter seems rather wasteful of resources (time and energy).

That chapter ended with a couple of examples of the methods used in the chapter. Note that these examples were designed to show the calculations that must be performed, thus driving home the steps and why those steps exist (hopefully). Statisticians use computers to perform these calculations for them.

In the next chapter, we move beyond trying to understand a single poll. Because so many polls are produced during election cycles, it becomes quite clear that analysts should not dismiss other polls without trying to use the information available. Thus, we will look at methods for reasonably combining polls into a better estimate.

4.5 Extensions

In this section, I provide some practice work on the concepts and skills of this chapter. Some of these have answers, while others do not.

Review

1. What information is needed to properly use stratified sampling?
2. How would you convince a person that an estimator with a lower MSE is preferred to one that is unbiased?
3. What is the source of bias in the stratified estimator?
4. What proportion of the US population agrees with the statement "Women should receive a year of paid maternity leave"?
5. What was it about the gender variable that made it desirable as the grouping variable?
6. What were some important conclusions from the stratified sampling experiments (pages 81–96)?
7. What was it about the first stratified sampling experiment (page 81) that produced results identical to the simple random sampling experiment?
8. In the formula for the stratified confidence interval, where did the term under the square root sign come from?

Conceptual Extensions

1. What is it about the "Democratic Support" map (Figure 4.1) that suggests polling should also take state into consideration? Why is state usually not taken into consideration in national polls?
2. In lieu of stratifying on the state, one could stratify on the region. Break the country into six good strata based on the state. Why is *your* stratification the best?

Chapter Appendix

3. Why is the mean squared error stratified estimator the same as that for the simple random sample estimator in the example on page 81?

Computational Extensions

1. Derive the formula for the confidence interval, Eqn. 4.3.

2. Determine if the simple random sample estimator has a smaller MSE than the stratified estimator under these conditions: Five strata with support $\pi_g = \{0.50, 0.30, 0.20, 0.90, 0.60\}$. The five groups are equally represented in the population and in your sample of 6000.

3. Determine if the simple random sample estimator has a smaller MSE than the stratified estimator under these conditions: Six strata with support $\pi_g = \{0.50, 0.30, 0.20, 0.90, 0.40, 0.70\}$. The six groups are equally represented in the population and in your sample of 6000.

4. Using the gender information from Table 4.2 for the actual voting results, estimate a 95% confidence interval for Clinton's support in the general population.

5. Using the party affiliation information from Table 4.2 for the actual voting results, estimate a 95% confidence interval for Clinton's support in the general population.

4.6 Chapter Appendix

4.6.1 R Functions

No statistician does these calculations by hand. We all use the computer. It helps to increase our precision and reduce the time needed to do the analysis. Here are some functions that help with the analyses covered in this chapter.

- stratifiedEstimator(w,s,n)

  This function calculates the stratified estimate and the margin of error. The inputs are the weights, w_g, the level of support in each sub-population, π_g, and the sample size for each group, n_g. For its code and an example of its use, please check out Section 4.6.3, below.

4.6.2 Two-Population Proportions Procedure

At points in the chapter, I asserted that women support the statement at a different rate than do men. As evidence, I presented sample data. In my sample, 420 of the 700 females polled supported the statement, but only 117 of the 300 males did. How can I state that these genders view the position differently *in the population*?

Statistics.

This section derives the statistical test for using the sample to determine if the support rate differs in the *population*. True, this is a bit overkill. All we need to do is believe that a difference exists. However, it does give some insight into how statistics moves forward.

The usual method for comparing two independent population proportions is to use the two-population proportions test. It is based on the Central Limit Theorem and the Normal approximation to the Binomial distribution.

The One-Population Case: Estimating π

For those who are interested in the theory, this section motivates the test, the test statistic, and the confidence interval when comparing the population proportions of two separate populations. To simplify the notation, it starts with a *one*-population proportion procedure. The two-population procedure is a simple extension and follows.

Recall from Section 3.1.2 (page 47) that the endpoints of the confidence interval are

$$\left(p - 1.96\sqrt{\frac{p(1-p)}{n}},\ p + 1.96\sqrt{\frac{p(1-p)}{n}}\right)$$

In that derivation, we approximated the Binomial distribution with a Normal distribution. Whenever approximating, it is important to determine how large of a sample size is needed to obtain *reasonably* accurate results.

Experiment: The following allows you to explore how large a sample size is needed to obtain a certain level of accuracy in the test. It relies on the proportion of the time the calculated confidence interval contains (or "covers") the true value of π. The coverage should be close to the claimed confidence level of 95%.

```
n  = 5              ## Sample size (change this)
B  = 1000000        ## Number of iterations (more is better)
pi = 0.500          ## True proportion
```

```
    x = rbinom(B, size=n, prob=pi)
    p = x/n

    lcl = p - 1.96 * sqrt( p*(1-p)/n )
    ucl = p + 1.96 * sqrt( p*(1-p)/n )

    sum( lcl<pi & pi<ucl )/B    ## Coverage (close to 95%??)
```

When I ran the above, I got a coverage rate of 93.7% when $n = 5$. This is a bit smaller than one should get (95%). This means that the confidence interval fails to cover π too often. As discussed in Chapter 3, the chapter dedicated to confidence intervals and p-values, failing to cover π enough indicates false precision in the estimates. That is: One will be more sure about them than one should be.

Increasing the value of n to 500 shows some improvement. The coverage becomes 94.5%. It is not until $n = 5000$ that coverage becomes 95.0%. Thus, a large sample is needed for the confidence intervals to be proper sized with respect to coverage.

◊ ◊ ◊

Close Enough?

With that being said, being off by 1pp may not be disastrous for estimation—unless the race is close. Thus, a sample size of 100 should be sufficient for your use if the candidates are not close in their support. That polls tend to have sample sizes in excess of $n = 500$ means that there is little error introduced by this approximation.

The Two-Population Case: Estimating $\pi_1 - \pi_2$

In the previous section, we derived the confidence interval for estimating a single population proportion, π. In this section, we obtain the confidence interval for the difference of proportions in two independent populations.

The test statistic is

$$Z = \frac{p_1 - p_2}{\sqrt{\frac{\pi_1(1-\pi_1)}{n_1} + \frac{\pi_2(1-\pi_2)}{n_2}}}$$

Here, p_1 and p_2 are the observed sample proportions, π_1 and π_2 are the population proportions, and n_1 and n_2 are the two sample sizes. This quantity follows a standard Normal distribution.

Deriving this follows the same steps as in the previous section, but uses the approximate distribution of the difference in proportions

$$P_1 - P_2 \overset{.}{\sim} \text{NORM}\left(\pi_1 - \pi_2, \frac{\pi_1(1-\pi_1)}{n_1} + \frac{\pi_2(1-\pi_2)}{n_2}\right)$$

Again, there is a π in the confidence interval (actually, a π_1 and a π_2). There are the usual two solutions: Substitute 0.50 for both π_1 and π_2 for a conservative interval or use p_1 for π_1 and p_2 for π_2 as in the Wald method.

Which to use? The usual rule of thumb is that if the value 0.50 is a reasonable value for π_1 and π_2, then you probably should use that option; otherwise, use the Wald method. But, whichever you use, make sure you are clear in your writing.

Be Very, Very Careful

Be aware that the two-population proportions test is only valid if you are estimating the proportions on *two separate populations*. It should **never** be used to compare two proportions in a single population.

In other words, using this procedure to compare candidate support by females to that by males is appropriate. The two populations are "female" and "male." They are separate populations.

However, using this procedure to compare support for two different candidates in a single population is *not* appropriate. The measurements are taken on a single population. In cases such as this, one should either focus on one candidate or use a Multinomial distribution (Chapter 7).

4.6.3 R Code for Stratified Sampling

This function calculates the stratified estimate and the margin of error. The inputs are the weights, w_g, the level of support in each sub-population, π_g, and the sample size for each group, n_g.

```
stratifiedEstimator = function(weight, support, n, conf.level
   =0.95) {
  alpha = 1-conf.level
  P = sum(weight*support)                   ## Total support

  E = -qnorm(alpha/2) * sqrt( sum(weight^2 * (0.25)/n) )
  return(c(support=P,margin=E))
}
```

Stratifying on Gender

To see how to use this, let us first calculate the confidence interval when stratifying on gender. First, specify the parameters of the sample according to the poll results (see page 87):

```
gender  = c(0.53, 0.47)
support = c(0.55, 0.43)
n = c(410,392)
```

Then, the code to run the function is

```
stratifiedEstimator(gender, support, n)
```

The output consists of two numbers, the estimate and the margin of error:

```
    support      margin
 0.49360000  0.03462877
```

Thus, a 95% confidence interval for Clinton support, when stratifying by gender, is 49.36% ± 3.46%, which is from 45.90% to 52.82%. Using a computer to perform the calculations makes life sooooo much easer.

Stratifying on Party

In addition to stratifying on gender, one can also stratify on the political party with the information available.

```
party   = c(0.34, 0.38, 0.28)
support = c(0.92, 0.42, 0.09)
n = c(267,298,231)

stratifiedEstimator(party, support, n)
```

This produces this output:

```
    support      margin
 0.49760000  0.03474332
```

Thus, a 95% confidence interval for Clinton support, when stratifying by party affiliation, is 49.76% ± 3.47%, which is from 46.29% to 53.23%.

4.6.4 Three Stratified Sampling Experiments

The purpose of these experiments is to explore what happens when the claimed weights do *not* match reality. Note that in the two experiments in the chapter (page 81), the stated weights did actually matched the population weights; that is, they were correct.

Estimates Far Off

This experiment explores what happens when the weights are wrong. The only change between this code listing and the one on page 81 is in the line marked <-. Compare that line to the pg line to see how much the weights are off from the real weights in the population.

```
### Population information
pi = c(0.95,0.35,0.25,0.45)    ## Subpopulation support
pg = c(0.25,0.25,0.25,0.25)    ## Real weights (population)
p = sum( pi*pg )               ## Population support

### Sample information
ng = c(250,250,250,250)        ## Subsample sizes
wg = c(0.45,0.15,0.20,0.20)    ## Claimed weights              <-

### Generate a million polls
x1 = rbinom(1e6, size=ng[1], prob=pi[1])
x2 = rbinom(1e6, size=ng[2], prob=pi[2])
x3 = rbinom(1e6, size=ng[3], prob=pi[3])
x4 = rbinom(1e6, size=ng[4], prob=pi[4])

### Stratified estimator:
estStS = wg[1]*x1/ng[1] + wg[2]*x2/ng[2] + wg[3]*x3/ng[3] +
    wg[4]*x4/ng[4]

### Mean squared error
mean( (estStS-p)^2 )
```

According to this experiment, the MSE is approximately 0.014 53. This is *much greater* than the MSE of the SRS estimator— more than 50 times higher. Thus, if the weights are this far from reality, the stratified estimator is much worse than the SRS estimator.

Note that the claimed weights differ from reality by as much as 20 percentage points. It is highly unlikely that a good analyst will miss the population proportions by that much. It is more likely that the differences will tend to be less than 5pp. The next experiment looks to see what happens when they are off by no more than 5 points.

Estimates Off by Quite a Bit

How bad is the MSE if the weights are all off by 5 percentage points in each stratum? To answer this question, substitute out the wg line with the following.

```
    wg = c(0.30,0.20,0.30,0.20)    ## Claimed weights          <-
```

Chapter Appendix 97

These results suggest that being off by this much makes the MSE *only* double that of simple random sampling. As such, being off by 5pp is still too much.

Estimates Off by a Tad

Clearly, the closer the weights are to reality, the lower the MSE of the stratified estimator. I leave it as an activity for you to check that being off by at most 2% in each group makes the MSEs roughly comparable.

In this case, the MSE for the stratified estimator is 0.000 23, a bit less than that of the SRS estimator. To obtain this, switch out the `wg` line with

```
    wg = c(0.27,0.23,0.27,0.23)   ## Claimed weights                  <-
```

5
The Bayesian Solution

Thus far, we have covered a lot of this beautiful discipline we call statistics. We have focused on estimators (and estimates) as well as confidence intervals and p-values. I hope I have made it clear that neither confidence intervals nor p-values tell us what we *want* them to tell us. In fact, for most people, they don't even tell us what we *think* they tell us.

Rectifying this issue requires us to look into an entirely different branch of statistics: Bayesian Analysis. While this was introduced in Chapter 3, this chapter delves much more deeply into it, exploring its strengths (many) and weaknesses (few).

Republic of Türkiye: The Parliamentary Election of 2023
Since the birth of the republic in 1923, one of the fundamental questions facing Türkiye was nationality. Atatürk believed that the only way to create a strong, stable Türkiye from the disintegrated Ottoman Empire was to unite all citizens into a single ethnicity. This did help with unifying Türkiye and protecting it from external incursion. Unfortunately, it also denied the existence of the Kurds in southeast.[1] They were referred to as "mountain Turks" and as "bandits." Their language and alphabet were restricted. Kurdish children had to have Turk names written in the Turkish language.[Boz21, SYR06] This oppression eventually led to the formation of the Kurdistan Workers Party (*Partiya Karkerên Kurdistanê*; PKK) in 1984. Arguably, this terrorist group is responsible for the deaths of 40,000 over these four decades. However, the vast majority of these deaths were of Kurds by the Turkish military, and not of Turks.[Eld19, Lus19]

In addition to the militant movement, Kurdish nationalism led to the formation of political parties. At that time, none were allowed to be explicitly

---

[1] According to the *Fondation-Institut kurde de Paris*, there are an estimated 20% of the Turkish population who identify as Kurds (based on 2016 estimates).[kdP17] Thus, they are not an insignificant minority.

The Bayesian Solution

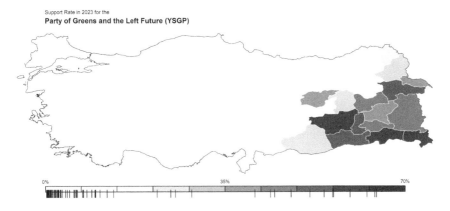

FIGURE 5.1
Map of the support for the Party of Greens and the Left Future (*Yeşiller ve Sol Gelecek Partisi*; YSGP) during the 2023 general election. These data are from the Supreme Election Council of Türkiye.[Sup23] Note that the upper limit on the scale is 70%. This choice helps to emphasize the regionalism of the party's support.

pro-Kurdish, and all campaign speeches had to be in Turkish. However, it is clear that certain political parties were supported by the Kurds. For instance, the People's Labour Party (*Halkın Emek Partisi*) was a pro-Kurdish political party in Türkiye, as were the Freedom and Democracy Party (*Özgürlük ve Demokrasi Partisi*, ÖZDEP), the Democracy Party (*Demokrasi Partisi*), the People's Democracy Party (*Halkın Demokrasi Partisi*), and several others.[oHR16] These parties were not contemporaneous. It seems as though that as one thrived, the government would link it to the PKK and outlaw it, and another pro-Kurdish party would arise to fill the political vacuum.[Wat09]

The current pro-Kurdish party is the "Party of Greens and the Left Future" (*Yeşiller ve Sol Gelecek Partisi*; YSGP). It formed in 2023 to contest that year's general election as a merger of the pro-Kurdish Peoples' Democratic Party (*Halkların Demokratik Partisi*; HDP) and the Green Left.[Duv23]. Unsurprisingly, the support for this new party is highly concentrated in the Kurdish regions of Türkiye (Figure 5.1).

Before the merger, the HDP was polling a consistent 10% in the 2023 general election. After the merger, the YSGP also polled a consistent 10%. From early April until early May, the average support for YSGP was a 9.9% with a standard deviation of only 0.88%.[Wi24c] Thus, given the consistency of polling for the YSGP at 10%, it seems unreasonable to ignore this additional information in estimating the current support for the YSGP.

For instance, on May 12 and 13, Özdemir conducted a poll of 3016 people.[Ara23] In analyzing the poll, should all the previous polls be ignored, or should that previous information be used to better their estimates?

This is an interesting question, for if we remain with frequentist methods, those methods we have been emphasizing since the start of the book, then previous polls must be entirely ignored. There is not even a mechanism to include previous polls in a frequentist analysis.[2]

However, if we think that the previous polls give some information about the current poll, and if we care about statistical legitimacy, then we need to move to Bayesian analysis. This chapter explores Bayesian analysis and its differences with frequentist analysis.

5.1 Overview: Why Bayesian Analysis

Your introductory statistics courses, whether in high school or college, tended to teach from the standpoint that the sample data are the source of all analyses. This is why so much time was spent on understanding the sample and checking that the sample met certain requirements. Once the data are understood, an appropriate procedure can be applied to the data to produce confidence intervals and p-values.

A result of this is that the confidence intervals (and p-values) did not measure what the researcher wanted. A 95% confidence interval gave a set of "reasonable" values for the population parameter. The p-value measured the probability of observing data this extreme—or more so—*assuming* that the null hypothesis is correct.

These frequentist, or "Fisherian," statistics are the foundation of much in introductory statistics. There are reasons in favor of it. First, the procedures are well understood and can be applied with little understanding of the distribution of the data.[3] Second, the calculations are relatively easy to calculate by hand (and all statistical programs have off-the-shelf functions to perform them if the sample is too large). Third, they do not need an advanced understanding of probability theory and distributions.

Unfortunately, all of this comes at a cost. That cost is that the confidence interval and the p-value are not measuring what we want them to measure.

---

[2]However, see Chapter 6 for a "cheating" way to combine polls. These methods are conceptually logical. However, they lack statistical rigor (if that matters).

[3]The Law of Large Numbers says that one can substitute the sample statistic for the population parameter if the sample size is large enough. For instance, one can use s^2 in the z-test instead of σ^2 if the sample size is large.

The Central Limit Theorem says that the distribution of the sample means will be approximately Normal if the sample size is large enough. For instance, if the data come from an Exponential distribution, then the sample means will be sufficiently Normal if the sample size is large.

Both of these mean that a large enough sample size is all that is needed to perform the standard statistical procedures (z-tests, etc.).

Overview: Why Bayesian Analysis 101

Fundamentally, we want both to be probability calculations. We want to be able to say that there is a 95% probability the confidence interval contains the population parameter. We want to say the p-value is the *probability* that the hypothesis (claim) is true.

They are not.

These weaknesses of Fisherian analysis are fixed in Bayesian analysis. When "doing Bayes," the result is an entire probability distribution. As a result, the "Bayesian confidence interval" (a.k.a. credible interval) is a probability statement. The probability that the parameter is in a 95% credible is, indeed, 95%. The Bayesian p-value is the probability that the hypothesis is true. For instance, if the claim is that the candidate is winning $\pi \geq 0.50$ and the Bayesian p-value is 0.33, then we conclude that the *probability* that the candidate is winning is 33%.

Bayesian analysis and Fisherian analysis are two fundamentally different approaches to statistical modeling and estimation. While both methods share the same goals of understanding and quantifying uncertainty, they differ in their underlying philosophy and methods.

5.1.1 Bayesian Advantages and Disadvantages

Here are some key advantages of Bayesian analysis over frequentist:

1. **Interpretability of results**: Bayesian analysis produces probability statements that are easily interpreted and intuitive, making it accessible to a wide range of audiences, including practitioners, policymakers, and stakeholders. In contrast, Fisherian methods rely on confidence intervals (and/or p-values), which are less straightforward to interpret and tend to lead to misconceptions and misinterpretations.

2. **Incorporation of prior knowledge**: Bayesian analysis allows analysts to incorporate prior knowledge or beliefs into their models through the use of probability distributions called priors. By doing so, Bayesian methods can better account for uncertainty and provide more informative estimates compared to Fisherian approaches that rely solely on the current sample data. That is, frequentists must ignore all previous analyses, but Bayesians can use that additional information.

3. **Coherent treatment of uncertainty**: Bayesian analysis provides a coherent framework for dealing with uncertainty by *updating* prior beliefs based on observed data to obtain posterior distributions. This approach ensures that all sources of information—including prior knowledge, expert opinion, and prior empirical evidence—are

treated consistently within the same probabilistic framework. Frequentists cannot do this.

Here are two more advantages of Bayesian analysis over frequentist. While these are more technical and not relevant to polling, they do provide more reasons for learning about Bayesian analysis.

4. **Computationally intensive methods**: Recent advances in computational algorithms—such as Markov Chain Monte Carlo (MCMC) methods—have made it easier to perform Bayesian analysis even when the likelihood function or posterior distribution lacks an analytical solution. This enables researchers to tackle complex and high-dimensional problems that may be difficult or impossible to address using frequentist techniques.

5. **Flexibility and adaptability**: Bayesian methods are highly flexible and can be tailored to specific applications or contexts by choosing appropriate prior distributions that reflect prior knowledge or beliefs. This contrasts with Fisherian approaches that rely on predetermined test statistics and significance levels, which may not always be suitable for a given problem.

While Bayesian analysis offers many advantages over Fisherian approaches, it also has a couple of limitations. These limitations (especially the first) are what made Ronald Fisher decide that Bayesian analysis was *not* an appropriate way of doing science. Some key drawbacks of Bayesian analysis include:

1. **Subjectivity**: Bayesian analysis requires specifying prior distributions to express prior knowledge or beliefs about the parameters of interest. This introduces subjectivity into the modeling process, as different analysts may choose different prior distributions based on their own opinions or preferences. While this can be a strength in some cases—such as when incorporating expert knowledge or historical information—it can also introduce bias or uncertainty if the choice of prior is not well-justified or appropriate for the problem at hand.

2. **Choice of prior distribution**: Selecting an appropriate prior distribution that accurately reflects prior knowledge or beliefs can be challenging, particularly in complex or high-dimensional problems where the choice of prior may significantly impact the results. Moreover, choosing a non-informative prior may not always be possible (or practical), as some prior distributions may introduce unwanted assumptions or constraints on the model.

3. **Computational burden**: Also not relevant to polling, but it bears mention. Although recent advances in computational algorithms

have made Bayesian analysis more feasible for many problems, it can still be computationally intensive and time-consuming, particularly for high-dimensional models or large data sets. This can make it difficult to perform sensitivity analyses or compare different model specifications, as each analysis may require significant computational resources.

Bayesian analysis offers several advantages over frequentist approaches by allowing analysts to incorporate prior knowledge, provide coherent treatment of uncertainty, make optimal decisions under uncertainty, flexibly tailor models to specific contexts, leverage computational advances, and produce interpretable results that are easily understood by a wide range of audiences. By leveraging these benefits, one can make more informed decisions under uncertainty and better address complex problems in various fields (such as election analysis).

However, while Bayesian analysis offers several advantages over Fisherian approaches, it also has some limitations. These limitations include subjectivity, difficulty in choosing prior distributions, and computational burden. By understanding these drawbacks, one can work to mitigate them to provide superior estimates of the population. An excellent resource for a deeper understanding of Bayesian Analysis is the book *Bayesian Data Analysis* by Gelman et al. It is currently in its third edition and is an awesome resource.[GCS+13] A second resource I recommend is be Albert's *Bayesian Computation with R*.[Alb09] Of course, if you are into the theory of Bayes, I find Robert's *The Bayesian Choice* to be a good, in-depth handling of all things Bayesian analysis from its roots.[Rob07]

The next section shows that some of these "drawbacks" can be successfully mitigated in practical research. We start with Bayes' Law and a toy example. Then we look at a shortcut to the mathematical calculations. With that shortcut, we get some more examples and an exposition on how Bayesian methods are used in the polling field, and how they *could* be used.

5.2 Bayes Law

In his posthumously published note to the Royal Society of London, the Reverend Thomas Bayes provides a definition of conditional probability as his Proposition 3.[Bay63, p378]

The probability that two subsequent events will both happen is a ratio compounded of the probability of the 1st, and the probability of the 2d on supposition the 1st happens.

Being one of the pioneers of probability theory, much of what Bayes wrote had to use words instead of well-understood symbols, because there were few well-understood symbols back then. To give this a *modern* look, what Bayes wrote is equivalent to

$$\mathbb{P}[\,A \cap B\,] = \mathbb{P}[\,A\,] \times \mathbb{P}[\,B \mid A\,] \tag{5.1}$$

The term on the left is read as "The probability of Event A *and* Event B happening" (the "intersection" of the two events). The second term on the right is read as "The probability of event B, *given* that event A occurs." It is also called the "**conditional probability** of B, given A" or "the probability of B, conditioned on A."

From this simple proposition, Bayes proves what is to become known as Bayes' Law. I am including its proof here because I enjoy showing how easily it is proven... when one uses modern symbols.

Theorem 5.1 (Bayes' Law). *Let A and B be possible events; that is, $\mathbb{P}[\,A\,] \neq 0$ and $\mathbb{P}[\,B\,] \neq 0$. Then,*

$$\mathbb{P}[\,A \mid B\,] = \frac{\mathbb{P}[\,B \mid A\,]\mathbb{P}[\,A\,]}{\mathbb{P}[\,B\,]}$$

That is, the probability of Event A, assuming Event B happens, is equal to the probability of Event B happening, given Event A happens, times the ratio of the probabilities of the two events.

Proof. This proof uses the definition of conditional probability in 5.1, the meaning of intersection, and basic algebra.

$$\mathbb{P}[\,A \cap B\,] = \mathbb{P}[\,A\,] \times \mathbb{P}[\,B \mid A\,] \quad \text{Proposition 5.1}$$
$$\mathbb{P}[\,B \cap A\,] = \mathbb{P}[\,B\,] \times \mathbb{P}[\,A \mid B\,] \quad \text{Proposition 5.1}$$
$$\mathbb{P}[\,A \cap B\,] = \mathbb{P}[\,B \cap A\,] \quad \text{Commutativity of intersection}$$

$$\mathbb{P}[\,A\,] \times \mathbb{P}[\,B \mid A\,] = \mathbb{P}[\,B\,] \times \mathbb{P}[\,A \mid B\,] \quad \text{Substitution}$$

$$\mathbb{P}[\,A \mid B\,] = \frac{\mathbb{P}[\,B \mid A\,]\mathbb{P}[\,A\,]}{\mathbb{P}[\,B\,]} \tag{5.2}$$

And it is proven.

Bayes Law

Moving symbols around may seem rather abstract, but proving this using just words is very cumbersome. As evidence of this, please see Bayes' note to the Royal Society of London, especially Section II and the scholium on page 392.[Bay63]

The next step is to see Bayes' Law in terms of probability *distributions* instead of merely as probabilities of single events. Doing this gives

$$g(\pi \mid \mathbf{X}) = \frac{\mathcal{L}(\mathbf{X} \mid \pi) f(\pi)}{\mathbb{P}[\mathbf{X}]} \qquad (5.3)$$

Compare Eqns. 5.2 and 5.3 to see the analogues. Since there are a lot of new symbols here, let's define them explicitly. After that, we will have a toy example to show how the parts fit together using Bayes' Law.

- $f(\pi)$

  This is the **prior distribution** of the parameter π. It contains the "prior" knowledge held by the researcher. The first step in any Bayesian analysis is quantifying the prior information about the parameter.

- $\mathcal{L}(\mathbf{X} \mid \pi)$

  This is the **likelihood function** of the observed data, $\mathbf{X}$. It contains the information that updates the prior distribution.

- $g(\pi \mid \mathbf{X})$

  This is the **posterior distribution** of the parameter π, given the new data, $\mathbf{X}$. It is the distribution of the parameter, π, given the new information. Knowing this distribution is the goal of Bayesian analysis because it contains all knowledge of the parameter after we analyzed the data.

- $\mathbb{P}[\mathbf{X}]$

  This is the probability of observing the data, $\mathbf{X}$. In my experience, it is a nuisance to deal with. Thankfully, there are ways of avoiding it entirely (Section 5.3). If we do not avoid it, then we will need to calculate

  $$\mathbb{P}[\mathbf{X}] = \int_0^1 \mathcal{L}(\mathbf{X} \mid \pi) f(\pi) \, d\pi$$

  for polling problems.

5.2.1 How is it Used?

So, now that we have Bayes's Law, what can we do with it? We can fix all of the weaknesses discussed in the previous chapter about confidence intervals and p-values. We can also use past information to improve our estimates, as this section shows.

"Toy" Example

Let us start with a toy example where the numbers work out nicely. Then, the next example will be more realistic. Here, let us estimate the proportion of my students who do the assigned reading, π.

The **first step** is to specify any previous information about the parameter π. Here, since we know we are estimating π, it must be bounded between 0 and 1. Beyond that, let's say there is no other prior information.

There are many distributions that meet this single requirement. The one that forces the least constraint on π is the standard Uniform distribution, $\text{UNIF}(0, 1)$. So, since this distribution best matches our knowledge, let us use it. Its probability function (a.k.a. its probability density function, pdf) over its domain is

$$f(\pi) = 1$$

The **second step** is to gather the data and quantify the knowledge contained in that data. When I gave them a quiz covering the readings, 5 of the 6 showed they had done the reading. So, I can say there are $x = 5$ successes and 1 failure in my class of size $n = 6$.

Since the number of successes follows a Binomial distribution (see Appendix B.2), the *likelihood* of the data is

$$\mathcal{L}(\mathbf{X} \mid \pi, n) = \binom{n}{x} \pi^x (1-\pi)^{n-x}$$

$$= \binom{6}{5} \pi^5 (1-\pi)^1$$

The **third step** is to calculate $\mathbb{P}[\,\mathbf{X}\,]$. Usually, there are methods to avoid this. I will visit them later, however. Here, let us directly calculate it.

$$\mathbb{P}[\,\mathbf{X}\,] = \int_0^1 \mathcal{L}(\mathbf{X} \mid \pi) f(\pi) \, d\pi$$

$$= \int_0^1 \binom{6}{5} \pi^5 (1-\pi)^1 \, 1 \, d\pi$$

$$= \binom{6}{5} \int_0^1 \pi^5 (1-\pi)^1 \, d\pi$$

$$= \frac{6!}{5! \, 1!} \int_0^1 \pi^5 (1-\pi)^1 \, d\pi$$

$$= \frac{6!}{5! \, 1!} \frac{6! \, 2!}{8!}$$

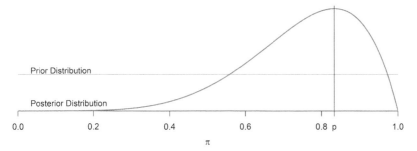

FIGURE 5.2
A comparison of the prior distribution, the observed proportion, and the posterior distribution of π given the data. Note that the observed data significantly changes the distribution of π from the prior (thin) to the posterior (thick). The vertical line marked "p" is the observed proportion, $5/6$.

To get the last line, I knew a trick.[4] Most people would use a computer to do the calculation. Note that this last *could* be simplified, but it actually makes future calculations easier to leave it as-is.

Finally, let us put everything together and calculate the probability density function for the posterior distribution:

$$g(\pi \mid \mathbf{X}) = \frac{\mathcal{L}(\mathbf{X} \mid \pi) f(\pi)}{\mathbb{P}[\mathbf{X}]}$$

$$= \frac{\frac{6!}{5!\,1!} \pi^5 (1-\pi)^1 \cdot 1}{\frac{6!}{5!\,1!} \cdot \frac{6!\,2!}{8!}}$$

Simplifying leads to

$$g(\pi \mid \mathbf{X}) = \frac{8!}{6!\,2!} \pi^5 (1-\pi)^1$$

From our knowledge of probability distributions, we recognize this last line as the probability function of a Beta distribution (Appendix B.4). Thus, the posterior distribution of π, given the new data, is

$$\pi \mid \mathbf{X} \sim \text{BETA}(6, 2)$$

Figure 5.2 shows the prior distribution, the observed proportion, and the posterior distribution. Note that the prior distribution is "flat" to indicate that we have no real information on π before the experiment. Also note that the posterior distribution is far from flat. This indicates we have more information about π.

---

[4] Thanks Mindy!! I've used that trick many, many, *many* times to avoid integration.

5.3 A Particularly Helpful Shortcut

Conjugate priors are a concept within Bayesian statistics, which allows one to easily calculate the posterior distribution given the likelihood function and the conjugate prior. Ultimately, the **"conjugate prior"** is a prior distribution which matches the likelihood function to produce a posterior distribution of the same form.

For instance, the prior distribution for the Binomial likelihood is the Beta distribution. This is because if the researcher uses the Beta as the prior distribution, then the posterior will also be a Beta. In fact, it can be shown that if the prior is $\pi \sim \text{BETA}(a, b)$, and the data consist of x successes and $n - x$ failures, then the posterior distribution is

$$\pi \mid \mathbf{X} \sim \text{BETA}(a + x,\ b + n - x)$$

The proof of this is in the appendix to this chapter (page 118). I do not think it adds anything to the discussion, so feel free to skip it—or not.

5.3.1 Using Conjugate Priors

To illustrate the use of conjugate priors, compare the form of these three quantities from the previous experiment:

- $f(\pi) = 1 = \frac{1!}{1!\ 1!}\ \pi^0 (1-\pi)^0$
- $\mathcal{L}(\mathbf{X} \mid \pi) = \frac{6!}{5!\ 1!}\ \pi^5 (1-\pi)^1$
- $g(\pi \mid \mathbf{X}) = \frac{8!}{6!\ 2!}\ \pi^5 (1-\pi)^1$

Notice that all three have the same basic form. Early in the days of Bayesian analysis, they noticed this. Then, they interpreted it. Finally, they formulated the theory of "conjugate priors."

These three quantities can be written in this form

- prior: $\pi \sim \text{BETA}(1,\ 1)$
- data: $x = 5, n - x = 1$ (successes $= 5$, failures $= 1$)
- posterior: $\pi \mid \mathbf{X} \sim \text{BETA}(6,\ 2)$

A Particularly Helpful Shortcut

And so, we can calculate the posterior distribution using very little math: specify the prior distribution as a BETA(a, b), collect the data (x successes and $n - x$ failures), write down the posterior as BETA$(a + x, \, b + n - x)$.

That really is all there is to it.

The interpretation is rather compelling, too. As discussed in the book's appendix (page 309), the two parameters of a Beta distribution can be interpreted as the amount of information known. The parameter a is the number of successes, and the parameter b is the number of failures. Together, they can be interpreted as a sample size, $n = a + b$.

Thus, for the "class preparation" example above, the prior distribution adds two "observations," 1 success and 1 failure. The data adds 5 successes and 1 failure. Therefore, the posterior distribution consists of 6 successes and 2 failures. In other words, selecting the flat prior we effectively added two virtual points to our data.

If we had selected a different prior distribution, then the posterior would be different. A good Bayesian will either explicitly defend the choice of a prior distribution, select a prior that has little effect on the data (as here), and/or investigate how robust the posterior distribution is to different possible prior distributions.

Experiment: Hearkening back to the introduction to this chapter, let us see how we can update our polling estimates. Between April 26 and May 3, Özdemir conducted a poll of $n = 5,916$ people. In that sample, 568 respondents stated that they would vote for the Party of Greens and the Left Future candidates (*Yeşiller ve Sol Gelecek Partisi*; YSGP).

Using the formulas from Section 3.1, we estimate the margin of error to be ± 1.3pp. Thus, we have a 95% confidence interval for YSGP support to be between 8.3% and 10.9%. Since this is a confidence interval, its interpretation is rather limited. We know that if we perform this sampling a gazillion times, then 95% of the calculated confidence intervals will contain the true support level for the YSGP, π.

Using Bayesian analysis (see Section 5.4 for the calculations), we can conclude:

> There is a 95% probability that YGSP support is between 8.9% and 10.4%.

Yes. With Bayesian analysis, we *can* make probability statements about the support level in the population, which is how we frequently *wrongly* interpret confidence intervals. *Also*, we are also not limited to standard claims. We can also conclude that there is a 15.7% probability that the YGSP support is at least 10%... or that there is a 5.2% probability that their support is at most 9%... or that there is a 79.0% probability that their support is between 9 and 10%.

Now, a bit over a week later (May 12 and 13), Özdemir held another poll. This time of $n = 3016$ people. In *that* sample, 335 responded that they would vote YSGP.[Ara23] We can use the previous information to update the estimated support level and *still* make probability statements using Bayesian analysis.

> There is a 95% probability that YGSP support is between 9.5% and 10.8%.

We can also conclude that there is a 64.2% probability that the YGSP support is at least 10%... or that there is a 0.01% probability that their support is at most 9%.... or that there is a 35.8% probability that their support is between 9% and 10%, which is a rather large decline due to the new poll.

Because we have the entire distribution, we can use it to answer *any and all* questions interesting to *us*.

◇ ◇ ◇

At one level, it is very comforting to be able to make probability statements about the party support. As a Fisherian, we are only able to make statements on our *confidence*. We can also use previous polls to better estimate party support. In Chapter 6, we discuss methods for combining polls to increase precision and accuracy in the estimates. Some of that is based on Bayesian methods, either explicitly or implicitly.

5.4 The Bayesian Calculations

The mathematics underlying the above Bayesian analysis is easy if we use conjugate priors (as I did here). So, let me show the math for the previous experiment here.

First, let us create a prior distribution for the April 26–May 3 poll. Since I know candidate support is bounded by 0 and 1, and since I have no additional information, I will use the flat prior.

$$\pi \sim \text{BETA}(1,1)$$

The Bayesian Calculations

The **second step** is to quantify the knowledge we learned from the poll. Here, we know that the number of people in the poll supporting YSGP follows a Binomial distribution. Additionally, from this particular poll, we know the number of successes is $x = 568$ and the sample size is $n = 5916$. Thus, the likelihood of the data is

$$\mathcal{L}(\pi \mid x, n) = \binom{n}{x} \pi^x (1-\pi)^{n-x}$$

$$= \binom{5916}{568} \pi^{568} (1-\pi)^{5916-568}$$

The **final step** is to calculate the posterior. Since we used a conjugate prior this calculation is as easy as adding.

$$\pi \mid \mathbf{X} \sim \text{BETA}(a+x,\ b+n-x)$$
$$\sim \text{BETA}(1+568,\ 1+5916-568)$$

Thus, the posterior distribution of π given the new data is

$$\pi \mid \mathbf{X} \sim \text{BETA}(569,\ 5349)$$

And, with that, we can perform our calculations using the computer.

- There is a 95% probability that the YGSP support is between 8.9% and 10.4%:
  ```
  qbeta(0.975, 569, 5349)
  qbeta(0.025, 569, 5349)
  ```

- We can also conclude that there is a 15.7% probability that the YGSP support is at least 10%...
  $\mathbb{P}[\pi \geq 0.10] =$ `1 - pbeta(0.10, 569, 5349)`

- ... or that there is a 5.2% probability that their support is at most 9%
  $\mathbb{P}[\pi \leq 0.09] =$ `pbeta(0.09, 569, 5349)`

Now that we have a better prior distribution based on the posterior calculated above, we can use it as the prior for analyzing the May 12 and 13 poll. That is, we have:

- Prior: $\pi \sim \text{BETA}(569,\ 5349)$

- Data: $x = 335$, $n = 3016$

- Posterior: $\pi \mid \mathbf{X} \sim \text{BETA}(569+335,\ 5349+3016-335)$

And so, the posterior distribution after the May 12 and 13 poll is

$$\pi \mid \mathbf{X} \sim \text{BETA}\,(904,\ 8030)$$

And, with that, we can perform our calculations using the computer.

- There is a 95% probability that the YGSP support is between 9.5% and 10.8%
  ```
  qbeta(0.975, 904, 8030)
  ```
  ```
  qbeta(0.025, 904, 8030)
  ```

- We can also conclude that there is a 64.2% probability that the YGSP support is at least 10%...

  $\mathbb{P}\big[\,\pi \geq 0.10\,\big] =$ `1 - pbeta(0.10, 904, 8030)`

- ... or that there is a 0.01% probability that their support is at most 9%.

  $\mathbb{P}\big[\,\pi \leq 0.09\,\big] =$ `pbeta(0.09, 904, 8030)`

Again, once we have the posterior distribution, we know everything possible about the parameter.

5.4.1 A Short Aside on Prior Distributions

There are a few things to take away from this series of experiments. The first is that the calculations are easy with the computer. They are also based on what *you* want to calculate.

The second thing to take away from this part is illustrated in Figure 5.3: Weak prior distributions (like the flat prior we used for the first Özdemir poll) influence the posterior distribution little, especially when there is a lot of data available (large sample size). Strong priors (like the one we used for the second Özdemir poll) influence it a lot.

This is seen in Figure 5.3 by the fact that the first observed support level for the YGSP party, p_1, goes through the highest point of the resulting posterior (i.e., it is the most likely support level), while the second observed support, p_2, only pulls the distribution in its direction (it is no longer the most likely support level).

The third thing to take away from this is subtle in the graphic, but clear in the analysis. As more and more data is taken the posterior distribution tends to become more and more narrow. This indicates that more data improves the precision of the estimates.

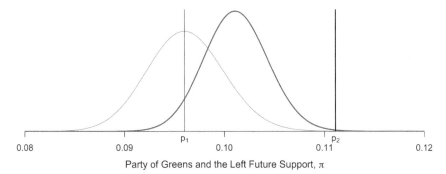

FIGURE 5.3
Comparison of the prior distribution (thin red curve) and the posterior distribution (thick blue curve) for the second Özdemir poll (May 12 and 13). The vertical lines are the observed support for the two polls. Note that the first distribution is heavily influenced by the observed data (as opposed to its prior). The second distribution is much less influenced by the observations because of the information in the prior.

5.4.2 A Second Short Aside on Prior Distributions

It may happen that you have prior knowledge about the support, but that you cannot put it into the above terms. For instance, you may be *very* confident that the support level for a party is between 15% and 19% (or 17% ± 2pp).

Assuming your "very confident" corresponds to a 95% level, you are stating that the expected value of π is 0.17 and the standard deviation is $\sigma = 0.01$. How would you quantify this information properly?

There are wonderful formulas to estimate α and β given the expected value and standard deviation. Because I use these formulas frequently, and because I want to do as little typing as possible, I created an R function to perform this calculation for me (and for you).

The function is betaPrior. It takes the expected value and the standard deviation. It returns the estimated value for α and for β. The actual function is provided in the appendix to the chapter (page 118).

Using the function, I estimate the two parameters for the Beta prior for the above example are

```
betaPrior(0.17,0.01)
```

The output

```
$alpha
[1] 239.7
```

```
$beta
[1] 1170.3
```

Thus, if I am very confident that the support level for a party is between 15% and 19%, my prior distribution would be

$$\pi \sim \text{BETA}(239.7, 1170.3)$$

There is more information about the Beta distribution in Appendix B.4. Personally, I find it very useful to model a single population proportion. However, it is not useful when modeling an election with more than two candidates. For information on how to model such an election, please refer to Chapter 7.

5.5 Bayesian p-Values

In polling, and in reporting polling results, one rarely comes across a hypothesis test or a p-value. The subject does not lend itself to hypothesis testing. Intervals are much, *much* more common. However, since we are here, and since the number of hypothesis tests out there is not zero, let me cover Bayesian p-values here.

The procedure for calculating Bayesian p-values starts the same as for calculating credible intervals. This is because credible intervals are probability statements just like Bayesian p-values are.

As with frequentists, we start with a claim about the population. However, different from frequentists, there is no need to create a null hypothesis and an alternative hypothesis.

Furthermore, one can directly calculate the *probability that the claim is true*. Frequentists will calculate the probability of observing data this extreme, or more so, assuming the null hypothesis is true. (Wow. What a mouthful.)

Thus, to show the procedure, let me return to the Toy Example on page 106. I claim that more than three-quarters of my students properly prepare for class. Thus, my claim is

$$H: \pi > 0.75$$

From my prior and the data, we determined that the posterior distribution was

$$\pi \mid \mathbf{X} \sim \text{BETA}(6,\ 2)$$

With that posterior distribution, we can directly calculate the probability that my claim is true:

- The probability that more than three-quarters of my students in this class properly prepare for class is 55.5%.

  ```
  1 - pbeta(0.75, 6, 2)
  ```

Notice that the interpretation is very straightforward. There is no need to "do the frequentist dance" and say

- The probability of observing data this extreme, or more so, assuming the null hypothesis ($\pi \leq 0.75$) is true, is 0.5339. Since this value is greater than $\alpha = 0.05$, we fail to reject the null hypothesis.

Again, Bayesian analysis gives a cleaner interpretation. In fact, I argue that it gives the interpretation that most think frequentist p-values gives. The only drawback is that one has to think more about the data and what processes generate it.

But, I think this should be the default action by statisticians.

5.6 Conclusion: Current and Future Use in Polling

As the above suggests, Bayesian analysis in polling is very useful. It is also rather straightforward and easily carried out. And yet, it sees little use in the industry. Sadness abounds.

There are some very good reasons to use Bayesian analysis (see Section 5.1). However, because introductory statistics is almost entirely Fisherian, Bayes tends to be relegated to the specialist.

There are two exceptions, however. First, many aggregating algorithms used to combine polls into a single estimate (see Chapter 6) tend to base their calculations on Bayesian analysis, whether purposely or not.

Second, there are a few journalistic outlets that attempt to use Bayesian analysis. They will report posterior estimated credible intervals. However, they will only use the flat prior, BETA(1, 1). This does allow them to state results in terms of probabilities. It also allows them to claim neutrality in that they are not using a strong prior to heavily affect the posterior distribution.

Personally, I find this position to be entirely understandable, but short-sighted. There is so much that Bayes can do to explore the results of many polls. In this chapter, we looked at using Bayes across polls. However, remember from Chapter 4 that one of the persistent issues is that stratified estimates are only good if important characteristics of the population are known. They are not.

However, one can use Bayesian methods to make explicit how different assumptions about the voters affect the vote result. This emphasizes the effect of the voter on the outcome of the election. It emphasizes that the election outcome is up to democracy.

5.7 Extensions

In this section, I provide some practice work on the concepts and skills of this chapter. Some of these have answers, while others do not.

Review

1. How does the of Figure 5.1 illustrate that the YGSP party appeals to Kurds?
2. How do frequentist methods differ from Bayesian?
3. What is the purpose of the prior distribution?
4. What is the purpose of the posterior distribution?
5. Why is it preferred to use a conjugate prior?
6. How does the interpretation of Bayesian intervals differ from frequentist?
7. How does the interpretation of Bayesian p-values differ from frequentist?
8. How is the inherent subjectivity of Bayesian methods mitigated?

Conceptual Extensions

1. Thinking back to the Toy Example on page 106, I sampled from a single, small class. Comment on problems with the sampling. Think about the purpose of a sample.
2. Again, thinking of the Toy Example on page 106, what is the probability that exactly 50% of my students prepare for class. Why is this answer 0%?

Chapter Appendix 117

3. Comment on the use of the BETA(1,1) prior in reporting poll results.

4. The BETA(1,1) prior effectively adds two data points (a success and a failure). Why don't we use BETA(0,0) as a prior distribution, thus adding no additional data?

Computational Extensions

1. According to the Toy Example on page 106, what is the probability that fewer than 10% of my students are properly prepared for class?

2. The proportion of the Turkish population that is Kurdish is reported to be 20% (reported by the *Fondation-Institut kurde de Paris* [kdP17]). Estimate each of the following using the non-informative (flat) prior.

    (a) What is the probability that the YSGP support exceeds that 20%?
    (b) What is the probability that the YSGP support is less than 10%?
    (c) What is a 80% credible interval for YSGP support?

3. Repeat the previous problem using BETA(100, 900) as the prior distribution.

5.8 Chapter Appendix

5.8.1 R Functions

These are R functions used in this chapter. Remember that no statistician will do complicated calculations by hand. We all use the computer to do the math work for us.

- `betaPrior(m, s)`

    This function calculates the best values of α and β, given your claimed values of m (expected value) and s (standard deviation). These calculated values are then used in the Beta prior distribution. The code for this function is given below on page 118.

- `pbeta(x, a, b)`

    This function calculates the cumulative probability of x from a BETA(a, b)

distribution. In others words, if $X \sim \text{BETA}(a, b)$, then this function calculates $\mathbb{P}\big[\, X \leq x \,\big]$.

- qbeta(p, a, b)

  This function calculates the pth quantile from a $\text{BETA}(a, b)$ distribution. In others words, if $X \sim \text{BETA}(a, b)$, then this function calculates x such that $\mathbb{P}\big[\, X \leq x \,\big] = p$.

5.8.1.1 Specific R Code Listing

The following is the code for estimating the parameters of a Beta distribution using the expected value and standard deviation.

```
betaPrior <- function(mu, sigma) {
    core = ( mu*(1-mu)/sigma^2 - 1 )
    a = core*mu
    b = core*(1-mu)
    return( list(alpha=a, beta=b) )
}
```

To use the function:

```
betaPrior(0.57, 0.01)
```

This is a simple function, but very useful.

5.8.2 Proof of the Conjugate Prior

This section shows that the Beta distribution is a conjugate prior for the Binomial likelihood. To do this, I first need to introduce the "kernel" of a probability distribution.

The Kernel

A probability distribution consists of two parts. The important part is the kernel. The kernel of a probability distribution is everything in the function that is a function of the random variable. For instance, in the Poisson distribution, whose probability function is

$$f(x) = \frac{\lambda^x \, e^{-\lambda}}{x!}$$

the kernel is

$$\frac{\lambda^x}{x!}$$

Chapter Appendix

The rest of the probability function is the normalizing constant $e^{-\lambda}$. It is only there to ensure that the probability over the sample space is 1.

The key is that if you work with kernels and come across $\lambda^x/x!$, then you know you are dealing with a Poisson distribution. There is no other option. Similarly, if you come across $\pi^a(1-\pi)^b$, when the random variable is π, you are dealing with a Beta distribution— guaranteed.

The Denominator

Now that we better understand the kernel of probability distribution, let us turn our attention to the denominator of Bayes' Law.

$$g(\pi \mid \mathbf{X}) = \frac{\mathcal{L}(\mathbf{X} \mid \pi) f(\pi)}{\mathbb{P}[\mathbf{X}]}$$

That denominator is just another normalizing constant. It is not a function of the random variable π. With this, we can rewrite Bayes' Law as the much simpler

$$g(\pi \mid \mathbf{X}) \propto \mathcal{L}(\mathbf{X} \mid \pi) f(\pi)$$

Here, because the denominator was a constant, we removed it and replaced the equality with "is proportional to" $\propto$. In fact, this technique is quite common in this field because we know that the normalizing constants will eventually multiply and divide to ensure the quantity on the right is a legitimate probability distribution.

The Proof

With that background on kernels of probability functions, let us see the easy proof.

Theorem 5.2. *The Beta distribution is the conjugate prior for a Binomial likelihood. In other words, if the likelihood is Binomial and the prior is Beta, then the posterior distribution is also Beta.*

Proof. Since we need to prove that if the likelihood is Binomial and the prior is Beta, then the posterior distribution is also Beta, let us start with a prior distribution and likelihood as written:

$$p(\pi) = B(\alpha, \beta) \, \pi^{\alpha-1}(1-\pi)^{\beta-1}$$

$$\mathcal{L}(\mathbf{X} \mid \pi) = \binom{n}{x} \pi^x(1-\pi)^{n-x}$$

The posterior distribution is calculated from the following

$$g(\pi \mid \mathbf{X}) \propto \mathcal{L}(\mathbf{X} \mid \pi) f(\pi)$$

$$= \binom{n}{x} \pi^x (1-\pi)^{n-x} \quad B(\alpha,\beta) \, \pi^{\alpha-1}(1-\pi)^{\beta-1}$$

Again, removing the constants, we have

$$g(\pi \mid \mathbf{X}) \propto \pi^x (1-\pi)^{n-x} \quad \pi^{\alpha-1}(1-\pi)^{\beta-1}$$

Combining like terms gives

$$g(\pi \mid \mathbf{X}) \propto \pi^{x+\alpha-1}(1-\pi)^{n-x+\beta-1}$$

Since this is the kernel of a Beta distribution, we know that the posterior follows a Beta distribution. Specifically, we know

$$\pi \mid \mathbf{X} \sim \text{BETA}(x+\alpha-1, \, n-x+\beta-1)$$

You are right. It does seem like the conclusion comes quickly. However, we are working with kernels. Thus, we just need to pay attention to the random variable, π.

■

6
Aggregating Polls

As frequentists, we have so far been drawing conclusions from a single poll. While this is a useful skill to have, elections are polled frequently; that is, there are multiple polls taken in a given election. Can we better estimate the population support, π, if we combine those multiple polls?

Clearly, since this chapter exists, the answer is yes. The more interesting question concerns how we can combine the polls *in a meaningful way*, which is the story of this chapter.

Republic of Korea: The Presidential Election of 2017
Since the start of the Sixth Republic in 1987, the Republic of Korea elects their President every five years in the middle of December. They use a national "first-past-the-post" system to elect the president to a single term. That is, the candidate who receives the most votes across the country wins the election.

In December 2012, Park Geun-hye of the center-right Saenuri party (previously named the Grand National Party, subsequently named the Liberty Korea party) was elected with 51.6% of the vote—the highest vote-share for president in the Sixth Republic. Her four years as president were not without scandal, and the South Korean National Assembly voted to impeach her in December 2016. This, along with the Constitutional Court's upholding of the impeachment, removed Park from office. This caused an early presidential election for her successor.

Figure 6.1 shows the support for each of the two second-place candidates, Ahn and Hong. Maps like these are interesting because they are able to give some insight to regional support of the candidates. For instance, Hong clearly does not poll well in the southwest, while Ahn seems to have consistent support throughout the country. This raises the question of why is Hong weak in that region (Gwangju city and South Jeolla province).

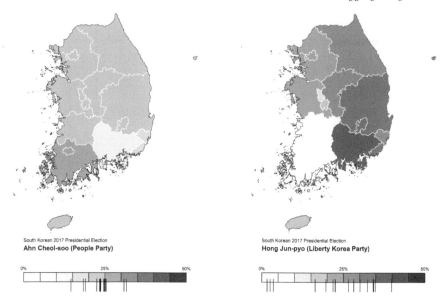

FIGURE 6.1
Map of the two-party support for the candidates vying for second place in the 2017 South Korean presidential election. Note that Ahn received 21.41% of the votes cast to Hong's 24.03%. The data are from Korea's National Election Commission.[Nat17]

In the last day before the polling blackout period, at least five polling firms conducted polls. All five of the polls had Moon leading his opponents by wide margins. However, they differed in their estimate of who would win *second* place, Ahn Cheol-soo or Hong Jun-pyo. The following were five of the final poll results for these two candidates, including the margin for Hong over Ahn.

Source		Ahn	Hong	Margin	n
Dong-A Ilbo	[Don17]	19.9%	17.7%	Hong −2.2	1058
Gallup	[Gal17]	20.0%	16.0%	Hong −4.0	1015
JTBC	[Han17]	19.7%	16.2%	Hong −3.5	1000
Metrix	[Mae17]	21.5%	16.7%	Hong −4.8	1500
Yeouido	[Yeo17]	20.1%	24.9%	Hong +4.8	2182

These five polling results tell very different stories about which candidate is in second place.

Yeouido estimates Hong will beat Anh by 2.0% to 7.6% (4.8±2.8%). Dong-A Ilbo and JTBC both conclude that the election is essentially a tie (the margin of error of ±3% is more than their difference). The other two polls conclude that Ahn will win second place (Gallup, 0.2% to 7.7%; Metrix, 1.7% to 7.9%). How can we reconcile these different estimates—and conclusions?

Perhaps it would behoove us to look at methods for combining the polls to increase estimation accuracy and precision. After all, each poll consists of a sample from the same population of interest. It would make sense to blend them in some meaningful manner.

6.1 Overview

In previous chapters, we looked at methods for estimating the current support for a candidate given a single poll. In this chapter, we examine some methods for combining polls. These methods range from simple averaging, to predicting future support, to complicated Bayesian methods. As usual, we will also look at the assumptions underlying these methods and the effects of misspecification.

Assumptions

Everything in this chapter makes the assumption that all polls are estimating the same **target population** (a.k.a. "population of interest"). This may seem like a reasonable assumption to make. However, because of how polling houses determine who is—and who is *not*—a "likely voter," the assumption is not entirely true in practice.

With that being said, the target populations *do* tend to be rather similar in practice. As such, the errors introduced tend to be rather minor. Note, however, that when two candidates are close, minor errors may be major enough to make the wrong prediction of the winner.

Oops.

Before we move on to the first method for combining polls, it is important that I review a few things from previous chapters. In a given poll, the number of people claiming to support a candidate depends on two independent parameters: the number of people polled (n, something we control) and the support level of the candidate in the population (π, what we are trying to estimate).

Since one is entirely determined by the polling firm and the other is entirely determined by reality, we know that adjusting one does not affect the other. This observation sits at the center of these attempts to aggregate polls.

Finally, to be clear, if the sample is randomly collected, then the number of people in the sample who support the candidate can be described as

$$X \sim \text{BINOM}(n, \pi)$$

For a refresher on the Binomial distribution, please visit Appendix B.2.

6.2 Simple Averaging of Polls

One option to combine polls is simply to average the estimates. Thus, for the five polls given in the introduction, we would add the individual estimates and divide by five to get an estimate for Hong's support to be 18.3%:

$$\text{Average} = \frac{17.7 + 16.0 + 16.2 + 16.7 + 24.9}{5} = \frac{91.5}{5} = 18.3\%$$

Piece of cake.

To obtain a confidence interval, however, we need to determine the correct sample size to use. There are two simple options: average and sum. Taking the average of the sample sizes gives $n = 1351$. Taking the sum of the sample sizes gives $n = 6755$.

This is a rather large difference with respect to the margins of error. The interval for $n = 1351$ is more than twice the width of the $n = 6755$ margin of error, as the following table indicates.

Method	n	Margin of Error
Averaged	1351	0.0267
Added	6755	0.0119

And so, from these results, can we decide which of these two methods is appropriate? In other words, which of these two methods provides a confidence interval[1] that contains the true value of π about 95% of the time?

To tell, let us do a statistical experiment.

---

[1] Remember that a confidence interval should contain the estimated population parameter a specified proportion of the time, when the sampling is done an infinite number of times. For a deeper refresher of what confidence intervals actually mean, please review Chapter 3, which is entirely dedicated to confidence intervals.

Simple Averaging of Polls

Usually, I keep the experiments in the chapter appendices. Here, because the result is so central to understanding how to determine the quality of a aggregation scheme, I include it in the prose. Note that the actual code is in the appendix, where it belongs (page 148).

Experiment: This experiment will generate poll results for five polls when $\pi = 0.183$, given the individual sample sizes from the polls on page 122. It will then calculate the resulting confidence interval based on the averaged sample size and second on the added sample size. Finally, in the final two lines, the code will determine how frequently the confidence intervals cover (contain) the value of π. It is this coverage rate that we hope will be close to 95%.

When I ran the code in the appendix (Section 6.9.4), I obtained the following values for the coverage.

Method	Actual Coverage	Claimed Coverage
Averaged	0.94	0.95
Added	0.55	0.95

What this means is that adding the sample sizes is a worse option than averaging the sample sizes. The confidence intervals calculated using the average of the sample sizes actually contains π only 94% of the time, which is a bit smaller than the claimed 95%, but rather close. For a statistician to be off by only a single percentage point is a day for celebration.

The confidence intervals based on adding the sample sizes, on the other hand, only contain π only 55% of the time. This is significantly less than our claim. In other words, in adding the sample sizes, we are over-stating how sure we are about the value of π.

◇ ◇ ◇

And so, according to this experiment, adding the sample sizes is not an appropriate method for combining polls. Averaging the sample sizes provides confidence intervals that are much more accurate (their coverage is closer to the claimed coverage).

6.3 Weighted Averaging of Polls

In the previous section, we developed a simple method for combining polls to obtain a confidence interval for the population support, π. While the accuracy of the confidence intervals using the averaged sample sizes is good, it would be nice to be able to increase the *precision* of the confidence interval. In other words, there may be a method for keeping the coverage at 95% (accuracy) but reduce the width of the confidence interval (precision).

A hint at how to proceed comes from the South Korean example above (page 122). Most of the polls had sample sizes around 1000. However, one had a sample size of $n_4 = 1500$, and another had a sample size of $n_5 = 2182$. Since the sample size is directly proportional to the precision of the estimate, the different sample sizes indicate different levels of information available in the poll. That is, the Yeouido poll contains twice as much information about the support level than does the Dong-A Ilbo poll. A better estimator should use this additional information.

The Sample *is* Information

Let me emphasize this again. As a frequentist, the sample is all you know about the population. Thus, the sample size is a measure of the information you have about that population.

More data is more information.

One method for doing this is to use a weighted average. The easiest way to visualize what is happening is to just put the polls together in terms of the number of people in favor and the number of people polled. For instance, 17.7% of the 1058 people in the Dong-A Ilbo poll (187 people) supported Hong, and 16.0% of the 1015 people in the Gallup poll (162 people) supported Hong.

Weighting is simply calculating the number of people in favor of the candidate in each poll and add those supporters together. The following table does this for the five polls from the introduction.

Source	Sample Size	Percent	People
Dong-A Ilbo	1058	17.7%	187
Gallup	1015	16.0%	162
JTBC	1000	16.2%	162
Metrix	1500	16.7%	251
Yeouido	2182	24.9%	543
Total	6755		1305

Weighted Averaging of Polls

Thus, the proportion of those in these five polls who support Hong is $1305/6755 = 19.32\%$. Using Eqn. 3.5 (page 51), a 95% confidence interval for Hong's support in the population is from 18.13% to 20.51%.

Note that this seems like an entirely natural way of combining polls. It is just treating each poll as a part of a larger "Mega-Poll" consisting of $n = 6755$ people, of whom $x = 1305$ support Hong. The fact that there are multiple polls involved increases the probability that an individual is selected by more than one poll. However, Chapter 2 shows that this is not a problem. So, at least from a logical standpoint, this seems like the best methods for combining polls. It takes a new problem and reduces it to one we previously solved.

Question: How does this estimator compare to the simple average estimator?

To answer this question, let us conduct a second experiment much like the previous one.

Experiment: Here, we will generate the polling results, calculate the weighted average, calculate a 95% confidence interval, and determine the proportion of those confidence intervals that contain the actual value of π (the coverage rate). We would hope about 95% of those confidence intervals contain π. The actual code can be found in the appendix on page 150.

Again, the last number is called the coverage rate of the procedure. It measures the proportion of the time that the confidence interval actually contains π. When I ran this code, I got an estimated coverage of 95%. Thus, like the one for the simple averaging, the confidence interval does not appear to be biased.

◇ ◇ ◇

So, which is better?

To determine which of the two unbiased confidence interval estimators is better, we need to look at the *precision* of each. All things being equal, we prefer the confidence interval that is narrower because it indicates that the estimate is more precise.

Adding these two lines to both experiments

```
mean( uprBoundAvg - lwrBoundAvg )
mean( uprBoundWtd - lwrBoundWtd )
```

gives us the average confidence interval width, a measure of precision.

The output tells the story and the *reason* we should use the weighted averaging method instead of the simple averaging method:

```
> mean( uprBoundAvg - lwrBoundAvg )
[1] 0.04657685
> mean( uprBoundWtd - lwrBoundWtd )
[1] 0.01843557
```

The average width of the simple-average confidence interval is 0.047, which is more than double the average width of the weighted-average interval, 0.018. Thus, the weighted-average interval is narrower than the simple-average interval. Since precision is a good thing, we should use weighted-average intervals.[2]

Arguably, the weighted-average interval *also* makes the most logical sense. The weights just recover the number of people in the poll who support that candidate. So, using weighted averaging is just like taking the multiple polls and treating them as part of one large Mega-Poll.

6.4 Averaging of Polls over Time

Thus far, we have looked at methods for combining polls taken on the same date. Usually, each election has multiple polls taken *over time*. For instance, the South Korean 2017 presidential election had dozens of polls in the five months leading up the election.[Wi20e] If we would like to increase the precision of our estimates, it would be good to include information from previous polls.

Clearly, we cannot just calculate a weighted average of all of the polls ignoring how long ago they were taken—or can we? Those polls taken five months ago may contain little current information. However, *if the support level is constant*, then early polls are just as useful as later polls; they all estimate the same target population.

This raises an important question: How can we tell if there is movement in support? Figure 6.2 plots the support for Hong in the last month of the election season according to each poll (dot). Note that his support level seems to be

---

[2] Note that this section looks at both accuracy and precision. Coverage speaks to accuracy, while interval width speaks to precision. As discussed in Section 2.2.3, low bias and low variance (a.k.a. high accuracy and high precision) are both important. In cases when one estimator has high accuracy and low precision, and the other has low accuracy and high precision, one can use the mean squared error (MSE) to determine the better of the two estimators. This is as true of estimators of a population parameter as it is of estimators of confidence intervals.

Averaging of Polls over Time 129

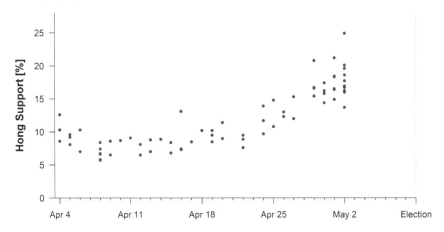

FIGURE 6.2
Plot of the estimated support level for Hong in the final month of the presidential campaign. The important take-away from this graphic is that Hong's estimated support level is not constant, which illustrates the importance of down-weighting old polls.

trending upwards. Thus, at least in this case, it would *not* make sense to use a weighted average of all of the polls, ignoring the time element. Furthermore, if the support level *does* vary through time:

- How do we determine which polls to use?
- How do we take time into consideration?
- If we wish to partially discount previous polls, how and what consistent rules should be used?

Of those questions, only one has an easy answer: How do we discount polls? Recall that in polling, the sample size is a measure of how much information the poll contains. Thus, if we reduce the sample size, then we are reducing the information the poll contains. This is especially good for two reasons: First, it is easily calculated. Second, changing the sample size only affects the amount of information available in the poll, it does not affect the estimated support.

The other questions do not have simple (any?) answers. For instance, how much should we discount a poll taken a week ago? Different rules will result in different estimates (and confidence intervals). And, because the research into this is scarce, we don't actually *know* the right answer (if such exists). It is an open question in polling studies.

Personally, I suspect the "right" answer depends on several factors, including the presence of important events (debates, scandals, etc.), the "stickiness" of the candidate in the population, and the candidate's fame. However, measuring this information and quantifying it to specify the weighting seems rather daunting.

6.4.1 Ignoring Time

The easiest solution is to just ignore time and combine all of the polls as in Section 6.3. The following R code loads the polling dataset from the local working directory, estimates the number of people in each poll supporting Hong, then forms the weighted average and calculates the confidence interval for his support using the binomial test.

```
korData = read.csv("kor2017pres-polls.csv")
attach(korData)

pHong   = Liberty.Korea    ## Hong support in each poll
nHong   = round( pHong*n ) ## People supporting Hong
allPoll = sum(n)           ## Sample size of the mega-poll
allHong = sum(nHong)       ## Hong support in the mega-poll

binom.test(allHong,allPoll)
```

The first two lines loads the data from the working directory and attaches its variables[3] The third line is a simple assignment line, making the code easier to read. The fourth estimates the number of Hong supporters in that particular poll.[4] The fourth and fifth lines calculate the total number of Hong supporters among all of the people asked in the polls.

The final line provides the 95% confidence interval using an exact procedure, rather than making the Normal approximation to the Binomial. If you would like to use the Normal approximation from Section 3.5, the number of successes (Hong supporters) is 11,956, and the number of trials (people asked) is 96,262.[5] This results in the estimated support for Hong to be between 12.2% and 12.6%. This is quite a bit below his actual vote share on election day, 24.03%.

---

[3]Technically, it appends (attaches) the variable names in the data set to the "variable path" along which R searches for variables. If this line is left off, the third line will provide an error. When using two data sets for an analysis, it may be undesirable to attach the data set, especially if the two data sets have variables with similar names. In such cases, the analyst will need to identify the variables using $ notation: korData$Liberty.Korean.

[4]If the raw data are provided, this step is not needed, as the actual number of Hong supporters would be known. When the sample size is sufficiently large, this estimation step adds very little to the bias of the estimated confidence intervals.

[5]Because of the large sample size, the 95% confidence interval you obtain will be from 12.2% to 12.6%. The difference between the Wald confidence interval and the binomial confidence interval occurs in the fifth decimal place. Similarly, the calculation done to estimate the number of Hong supporters matters little when the poll size is large, like here.

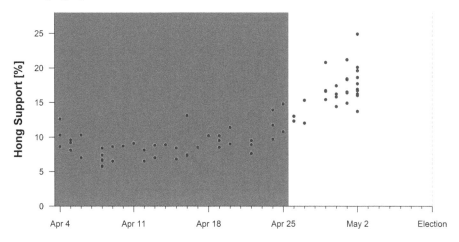

FIGURE 6.3
Plot of the estimated support level for Hong in the final two months of the presidential campaign. The *un*shaded region contains the subset of polls used in Section 6.4.2.

6.4.2 Ignoring Old Polls

A second option is to ignore polls older than a specified number of days. The code is a bit more difficult, but only because we are limiting the number of polls being examined with [doy>114], the "day of the year is greater than 114" (April 24). Assuming the code in Section 6.4.1 is run, the code to estimate Hong's support based solely on the last week of polls is

```
allPoll = sum(n[doy>114])
allHong = sum(nHong[doy>114])

binom.test(allHong,allPoll)
```

When using a weighted average of all of the polls after April 24—the last two weeks —we are 95% confident that the support for Hong on May 2 is between 16.5% and 17.2%.

Note that this confidence interval is very different from the previous one. The plot of Hong support over time suggests why (Figure 6.3). His support significantly increased over the last couple of weeks of polling.[6] Thus, the method of Section 6.4.1 will include polls from a time when Hong's support was much different than those at the end.

---

[6]South Korea implements a blackout period on reporting poll results during the final week of the campaign (I am jealous). This is the reason for the lack of polls between May 3 and May 9, election day.

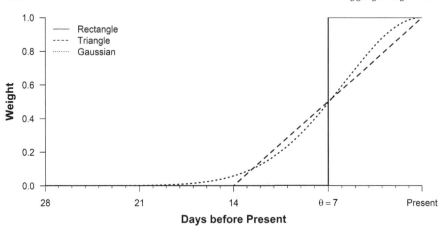

FIGURE 6.4
A plot of three weighting functions described in the text. Each is scaled here to weight a one-week-old poll 50% of a new poll.

6.4.3 More Sophisticated Poll Weighting

The previous section only includes polls from the past week and fully ignores those before. This raises the question of why the separation should be so sudden. Is a poll from 8 days ago really worth 0% of a poll from just 7 days ago? Should that one day make such an incredible difference in the information from the poll? Instead of such a dramatic change in weighting, it may be more reasonable to weight polls using a smoother weighting function. Figure 6.4 shows a few weighting functions that can be used.

The weighting function implemented previously in Section 6.4.2 is the rectangular function (a.k.a. the Heaviside step function). The triangular function reduces the weight on the poll linearly as time passes. The Gaussian function weights the more-recent polls more and the not-so-recent less than the triangular weighting, but gives a positive weight to all polls. (All three functions are shown in Section 6.9.1, page 144.) The rectangular weighting width can be altered according to one's expectation on the importance of past polls, as can that of the other two. This parameter also needs to be set by the researcher based on an understanding of how quickly candidate support changes in the population.

To illustrate this, let us use the triangle weighting with a half-weighting of one week (at one week, the poll is weighted by 50%). Figure 6.5 illustrates the weights given to the polls.

Averaging of Polls over Time

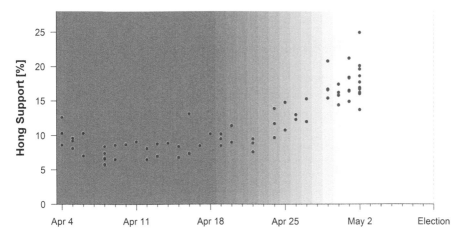

FIGURE 6.5
Plot of the estimated support level for Hong in the final two months of the presidential campaign. The background color shows the triangular weighting ($\theta = 7$, one week) given to the polls. The darker the background, the lower the weight.

The following is the R code I used to estimate Hong's support on May 2 (the 122nd day of the year). To make things a bit easier to understand, I created the weighting function listed in the appendix to this chapter (page 145). With that function, the following code weights all of the polls for the past two weeks ($\theta = 7$ days) and calculates a 95% confidence interval for Hong support on May 2.

```
allPoll  = triangleWeight(122-doy,7,n)
allHong  = pHong*allPolls

N = ( sum(allPoll) )
X = ( sum(allHong) )

wald.test(X,N)
```

According to this weighting scheme and the Wald procedure, a 95% confidence interval for Hong's support on May 2 is from 16.2% to 17.0%.

To illustrate the effect of the weighting function and the theta parameter, Table 6.1 compares the confidence intervals for several combinations of weighting functions and theta values. Note that changing the weighting function does not affect the confidence interval as much as does changing the parameter θ. This is because Hong's support increased dramatically in the final couple of weeks of the campaign (see Figure 6.2). Thus, it is more important to correctly specify the polls to include than it is to correctly weight them.

TABLE 6.1
Table showing how the confidence interval estimates for Moon vary according to the weighting function used.

Function	Theta	Lower Bound	Upper Bound
All		12.2	12.6
Rectangle	7	16.5	17.2
Triangle	7	16.2	17.0
Gaussian	7	16.2	16.9
Rectangle	14	14.8	15.4
Triangle	14	14.4	15.0
Gaussian	14	14.4	15.0

6.5 Looking Ahead

In the previous section, we looked at how to estimate *current* candidate support. This is a useful skill to have. However, it may be more important to be able to predict *future* candidate support. While this may be very important in settings where there is a period of time before the election where polls are not allowed to be published, it is also important in cases where one is close to the election and wants to predict the results.

Assumptions

In order to draw meaningful conclusions from the procedures in this section, one must assume that the trend continues into the future. This may be a safe assumption over the short-term. However, whether it holds long-term will depend on the idiosyncrasies of the particular election (nd election cycle).

If a major event occurs between the current poll and the election, then the quality of the estimate suffers.

To do this, we need to think more about our problem. Figure 6.6 provides the polls, along with the actual result of the election. As is frequently the case, the graphic points us in the right direction. From April 22 until the start of the polling blackout, it appears as though Hong's support consistently increases. It would make sense to take this information into consideration when estimating his support on election day (May 9). We have a common method for doing this: linear regression.

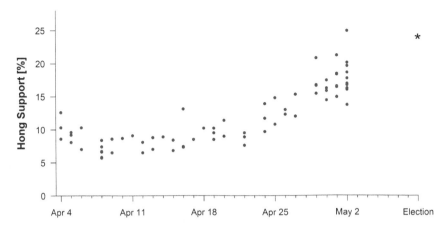

FIGURE 6.6
Plot of the estimated support level for Hong in the final two months of the presidential campaign. Note that his estimated support level is not constant. The star on election day is Hong's actual vote share (24.03%).

6.5.1 Regression

In its simplest form, linear regression seeks to summarize the linear relationship in the data with a single line. The strength to this is that its statistical properties are well understood and familiar with those who have taken an introductory statistics course. The data model is

$$y_i = \beta_0 + \beta_1 x_i + \varepsilon_i$$

Here, y_i is the candidate support rate (dependent variable), x_i is the date (independent variable), and ε_i is the residual (how far off the observed value is from the estimated value, the line). Furthermore, β_0 is the y-intercept, the value of y when $x = 0$, and β_1 is the slope, the amount y increases as x increases by 1 day.

The purpose of regression is to properly estimate β_0 and β_1. The standard method to do this is by using ordinary least squares (OLS). This is the usual method taught. Using calculus, one can show that the estimator of the slope is

$$b_1 = \frac{\sum_{i=1}^{n}(x_i - \bar{x})(y_i - \bar{y})}{\sum_{i=1}^{n}(x_i - \bar{x})^2}$$

This result is proven in this chapter's appendix, Section 6.9.

These estimators have a couple of good properties. First, they are unbiased (recall Section 2.2.1). Second, and more important, they have the lowest MSE of any other unbiased estimator (recall Section 2.2.3). Unfortunately, these results only hold *if* the relationship between the two variables (candidate support and time) is linear *and* if the residuals ε_i have a constant variance.

The first requirement is usually not a major problem, assuming that there is a period through which a line will fit nicely and assuming that nothing major happens in the race between then and the election.

The second requirement is frequently problematic in this use. Recall that the dependent variable is a proportion with expected value π, and that the variance of a proportion is $\pi(1-\pi)/n$ (Section 2.2.3). Since the variance is a function of the expected value of the dependent variable, the residuals will have variances that are also a function of the dependent variable. Hence, a violation of one of the OLS requirements.

OLS Alternatives: Weighting

A fix is rather easily made. Instead of using ordinary least squares, one can use *weighted* least squares (WLS), where the data points are weighted according to the sample size. Those polls with greater sample sizes are weighted higher than those with smaller sample sizes. The mathematics underlying this method do fix the inherent heteroskedasticity (non-constant variance).

Beyond the mathematics, there is a natural appeal to emphasizing larger polls over smaller ones. Polls with larger sample sizes contain more information (the sample *is* the information) and should contribute more to the estimate than smaller polls. This philosophy reflects the weighed averaging of Section 6.3.

In R, the code is

```
allPoll = triangleWeight(122-doy,7,n)
allHong = pHong*allPoll

korModWLS = lm(pHong~doy, weights=allPoll)
summary(korModWLS)
```

The first two lines weight the polls according to their age (recall Section 6.4.3). The third line does the weighted least squares modeling, weighting on the size of the poll after adjusting for age. Finally, the fourth line provides the regression table

```
              Estimate   Std. Error   t value    p value
(Intercept)   -0.829604    0.153781    -5.395    3.13e-06
doy            0.008310    0.001283     6.476    9.11e-08
```

Looking Ahead

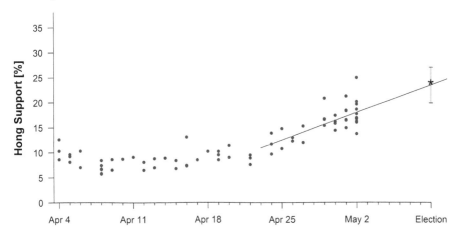

FIGURE 6.7
Plot of poll results for Hong in the final month of the presidential campaign. The line is the WLS regression line. The interval corresponds to the WLS 95% confidence interval. The plus is the support received by Hong on election day.

This indicates that the equation for the line of best fit is

$$y = -0.829\,604 + 0.008\,310x$$

In other words, for each additional day in the campaign (doy increasing by 1), Hong's support tends to increase by 0.8310 percentage points.

Since we are merely trying to determine the line of best fit for the data, and not testing any hypotheses, the other information in the regression table output is of limited value to us and can be safely ignored.

The straight black line in Figure 6.7 is this line of best fit using weighted least squares (WLS). The interval to the right of the plus is the 95% confidence interval for Hong's expected vote share. That the actual vote share is included in this interval, and no others yet, is strong evidence that forecasting adds accuracy to estimating the actual voter support on election day.

To obtain the confidence interval, run

```
predict(korModWLS, list(doy=129), interval="confidence")
```

The output is

```
        fit       lwr       upr
1 0.2423921 0.217333 0.2674513
```

This output means that the estimated level of Hong support on May 9 is 24.24%, with a 95% confidence interval from 21.7% to 26.7%. That Hong's actual vote share was 24.03% suggests that including regression to estimate future vote results is a fantastic idea.

Or, maybe we just got lucky.

OLS Alternatives: Beyond the Ordinary

Weighted least squares is not necessarily the best option. It still assumes that the residuals follow a Normal distribution. Strictly speaking, this is not the case— *cannot* be the case. To address the fact that the dependent variable is discrete, we can model it either as a Binomial or a Beta-Binomial random variable. The former requires introducing generalized linear models (GLMs); the latter, vector generalized linear models (VGLMs). My experience suggests that this is a promising direction the explore. However, the improvement in accuracy is quite minor. In other words, there is no *practically significant* gain in using these more-complicated regression methods.

With that being said, I encourage you to read about the uses and applications of both GLMs and VGLMs. They are fascinating advancements in regression. Here are some sources: [MT11, Woo06, Yee98, Yee15].

6.6 South Korean 2017 Presidential Election

To see how to use these methods, let us estimate the vote share for the other two main candidates in the 2017 South Korean presidential election. The data can be found in the `kor017pres-polls.csv` file.

6.6.1 Moon Jae-in

The candidate for the Democratic Party in the election was Moon Jae-in. He was the Democratic candidate in the 2012 election, but lost to Park Geun-hye. The polling results for the last month in the election cycle are provided in Figure 6.8. Notice how stable his support is throughout the entire month. Thus, it would seem that regression would not be needed. However, let us calculate all of the confidence intervals for Moon.

The results are given in Table 6.2. Note that there is little difference in the confidence intervals, regardless of which weighting function is used. This is a result of the stability of Moon's support throughout the final month of the campaign. All of the confidence intervals miss the election result of 41.08%,

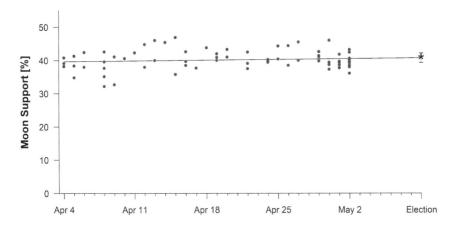

FIGURE 6.8
Plot of poll results for Moon in the final month of the presidential campaign. The plus is the actual support received by Moon on election day.

however. This is not entirely surprising, since these confidence intervals are attempting to estimate his support on May 2, not on the election day.

Since Moon's support over the past month was relatively stable, one would not expect much benefit from using regression. Regardless, let us use it to get an estimate for his vote share in the election. The 95% confidence interval is from 39.5% to 42.5%, with a point estimate of 40.96%.[7] This also tells a similar story as the weighted averages in Table 6.2, albeit the interval is wider because it is estimating a future event. The intervals of Table 6.2 are estimating Moon's support on May 2, the last day polls were published.

On the other hand, the regression estimates are actually for May 9, the day of the election. Because it is estimating a future event, the confidence interval is wider than for a present event. Note that Moon's actual support in the election was 41.08%, which is well within the confidence interval.

Here is the code used to obtain these results. The code for the weighting functions can be found in this chapter's code appendix, Section 6.9.

```
##### Load the data
korData = read.csv("kor2017pres-polls.csv")

# Obtain important variables without attaching
pMoon = korData$Democratic
n     = korData$n
doy   = korData$doy
```

---

[7]For this estimate, I decided to use all of the data. This decision came about because there was no significant change in Moon's support over the last month of the campaign.

TABLE 6.2
Table showing how the confidence interval estimates for Moon vary according to the weighting function used. They do only slightly.

Function	Theta	Lower Bound	Upper Bound
All		39.8	40.4
Rectangle	7	40.0	40.9
Triangle	7	39.8	40.7
Gaussian	7	39.8	40.7
Rectangle	14	40.1	40.9
Triangle	14	40.0	40.8
Gaussian	14	40.0	40.8

```
##### Create the weighted table
#    Uncomment the weighting scheme you wish to use.
allPoll = rectangleWeight(122-doy,1000,n)
#allPoll = rectangleWeight(122-doy,7,n)
#allPoll = rectangleWeight(122-doy,14,n)
#allPoll = triangleWeight(122-doy,7,n)
#allPoll = triangleWeight(122-doy,14,n)
#allPoll = gaussianWeight(122-doy,7,n)
#allPoll = gaussianWeight(122-doy,14,n)

# Now, calculate the confidence interval
allMoon = pMoon*allPoll
N = round( sum(allPoll) )
X = round( sum(allMoon) )
binom.test(X,N)

##### Confidence interval results from using
#    WLS regression
allPoll = rectangleWeight(122-doy,1000,n)
allMoon = pMoon*allPoll

korModLmMoon = lm(pMoon~doy, weights=allPoll)
predict( korModLmMoon, data.frame(doy=129),
   interval="confidence" )
```

Note that the first section of the code loads the data and only make available the three variables used in the analysis. If you would rather not attach the data, this is the preferred method to access the variables.

The second section does the weighting of the poll sizes and calculates the confidence interval. The code includes seven weighting schemes, but only the first is left un-commented. This practice allows the analyst to select the weight function without having to repeat a lot of code.

South Korean 2017 Presidential Election 141

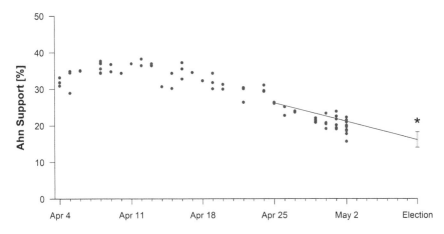

FIGURE 6.9
Plot of poll results for Ahn in the final month of the presidential campaign. The plus is the actual support received by Ahn on election day.

Finally, the third section calculates the confidence interval for Moon's support on election day (doy = 129) using all of the data. The output from this fourth section is

```
          fit       lwr       upr
1   0.4096053 0.3946944 0.4245161
```

This indicates that the expected vote outcome is 40.96%, with a 95% confidence interval from 39.47% to 42.45%.

6.6.2 Ahn Cheol-soo

In the case of Moon and Hong, regression did a superior job of estimating the vote share earned on election day. Let us turn our attention to estimating Ahn's vote share. Figure 6.9 shows poll results over the past month for his candidacy. Note the severe downward trend over the past few weeks. That it lasted so long suggests that the trend will continue. In fact, the 95% confidence interval using regression provides bounds of 11.6% to 14.0% (using the Gaussian weighting function with $\theta = 7$).[8] His actual vote share on election day was 21.41%, which is far outside the interval.[9]

---

[8] I selected this weighting function because it seemed as though Ahn's decline may have been ending. The Gaussian weighting function provides a higher weight to young polls than does the triangle function. This is evident from Figure 6.4, where the Gaussian function is higher than the triangle function for days closer to the present.

[9] The code to perform this analysis can be found in this chapter's code appendix, Section 6.9.

Thus, regression is not a panacea for estimating vote share in the future. Care must be taken, especially with the assumption that nothing significant happens between the last poll and election day; i.e., that current trends continue. There is nothing in the known data that suggests a change in Ahn's decline in the polls.

However, it could be that Ahn had a strong base of support that time would not (or could not) wear away. And, it is this base that stayed with him in the election. This phenomenon is quite common among politicians. There is a group of people that are die-hard supporters. Being able to estimate that proportion would help determining how much of the population really is "in play" in the election.

Something to think about.

6.7 Conclusion

This chapter looked at how to combine several polls to improve the accuracy and the precision of our estimates. It started with the simplest of averaging schemes: simple averaging of the support level. From that simple beginning, we progressed to estimating the election outcome using linear regression. Along the way, we thought more deeply about what it means to combine polls together and why accuracy *and* precision are so important to estimation. Of course, we still needed to remember the assumptions being made and to evaluate whether they were reasonable given the realities of the election season.

The final results suggested that weighted least squares regression offers a good estimate of candidate support for the near future. For the 2017 South Korean presidential election, it was only the weighted least squares method that successfully estimated the outcome for Hong. It failed in predicting the final support for Anh.

However, these frequentist methods all suffer from the same flaw: there is no way to provide probabilities or odds of one candidate winning over another. Frequentist methods only allow us to conclude that a set of outcomes is "reasonable," given the data and our definition of reasonable. Bayesian methods do not have this flaw. Such methods allow us to make statements about the probability of the candidate winning, or of the candidate getting less than 20% of the vote, etc.

In the next chapter, we provide an extended example of the techniques we have covered in these first three chapters, along with some short extensions. Hopefully, this will illustrate the fundamental difficulties in predicting an election outcome, even when the data is there. All polling relies on two assumptions that may not be true: the respondents are telling the truth, *and* the respondents are representative of the population of interest.

6.8 Extensions

In this section, I provide some practice work on the concepts and skills of this chapter. Some of these have answers, while others do not.

Review

1. Our first attempts at averaging polls was to simply average the proportions. In that case, how did we determine that averaging the sample sizes was better than adding them?
2. Why does a weighted average make more sense than a simple average?
3. What is the difference between accuracy and precision?
4. How did we measure accuracy in Section 6.2?
5. How did we measure precision in Section 6.3?
6. Why is it important to take into consideration the age of the poll?
7. Which of the three weighting functions described in the text is the best? How would we know?
8. Why did we use linear regression in this chapter? When would it be appropriate to use? When would it *not* be appropriate?

Conceptual Extensions

1. Using just the maps of Figure 6.1, can you tell which candidate is ahead? What additional information would you need? Why are maps like these misleading for determining which candidate is ahead?
2. The basic assumption for this chapter (page 123) is that the polls are estimating the same population. Why does this assumption need to be stated?
3. What is a "likely voter," and how would you determine one?
4. Why is it more satisfying to weight polls with larger sample sizes more than those with smaller?

5. Why do we expect the coverage rates in Section 6.2 to be close to 95%?
6. What would happen to the interval widths if the confidence level is changed from 95% to 90%? What would happen if we change them to 100%?
7. How should we determine the appropriate weighting function (Figure 6.4)?
8. Why were the estimates so good for Moon and Hong, but so wrong for Ahn?

Computational Extensions

1. Estimate the Hong support level on election day using other weighting functions. Which of the weighting functions of Table 6.1 produces the best estimate of his actual support level?
2. Devise an experiment to determine when the Wald procedure is superior to the exact Binomial procedure in estimating the population proportion.

6.9 Chapter Appendix

6.9.1 Three Weighting Functions

Here are the three weighting functions discussed in this chapters, specifically in Section 6.4.3. These are used to weight polls according to when they were performed.

Rectangular (or Heaviside)

```
rectangleWeight <- function(daysAgo, theta, n) {
  weight = ifelse(daysAgo>theta, 0, 1 )
  return(n*weight)
}
```

In this, daysAgo is the number of days ago the poll was taken, theta is the half-weight of the function (time when the poll is weighted 50%), and n is the sample size of the poll. The ifelse function tests the provided statement (daysAgo>theta) and sets a 0 weight if the statement is true or 1 if it is false. Finally, the function returns the effective (weighted) sample size.

Chapter Appendix

Triangular

```
triangleWeight <- function(daysAgo, theta, n) {
  weight = ifelse(daysAgo>2*theta, 0, (1-daysAgo/(2*theta)) )
return(n*weight)
}
```

As above, daysAgo is the number of days ago the poll was taken, theta is the half-weight of the function (time when the poll is weighted 50%), and n is the sample size of the poll.

The ifelse function tests the provided statement (daysAgo>2*theta) and sets a 0 weight if the statement is true or a (1-daysAgo/(2*theta)) if it is false. Again, the function returns the effective (weighted) sample size.

Gaussian

```
gaussianWeight <- function(daysAgo, theta, n) {
  weight = dnorm(daysAgo/theta*1.17741)/dnorm(0)
return(n*weight)
}
```

As above, daysAgo is the number of days ago the poll was taken, theta is the half-weight of the function (time when the poll is weighted 50%), and n is the sample size of the poll.

There is no need for an ifelse function because all polls receive a positive weight. The form of the Gaussian-weighting function is similar to that of the previous ones. The differences in the second line arise from ensuring the weight today (daysAgo = 0) is 1 and that the weight at daysAgo = theta is one-half.

This function also returns the effective (weighted) sample size.

6.9.2 Two Analyses

Ahn Analysis

This contains the code to perform the analysis of Ahn for Section 6.6.2.

```
##### Load the data
korData = read.csv("kor2017pres-polls.csv")

# Obtain important variables without attaching
pAhn = korData$People.s
n    = korData$n
doy  = korData$doy

##### Create the weighted table
#    Uncomment the weighting scheme you wish to use.
allPoll = rectangleWeight(122-doy,1000,n)
#allPoll = rectangleWeight(122-doy,7,n)
#allPoll = rectangleWeight(122-doy,14,n)
```

```
#allPoll = triangleWeight(122-doy,7,n)
#allPoll = triangleWeight(122-doy,14,n)
#allPoll = gaussianWeight(122-doy,7,n)
#allPoll = gaussianWeight(122-doy,14,n)

# Now, calculate the confidence interval
allAhn = pAhn*allPoll
N = round( sum(allPoll) )
X = round( sum(allAhn) )
binom.test(X,N)

##### Confidence interval results from using
#    WLS regression
allPoll = triangleWeight(122-doy,7,n)
allMoon = pAhn*allPoll

korModLmAhn = lm(pAhn~doy, weight=allPoll)
predict( korModLmAhn, data.frame(doy=129),
   interval="confidence" )
```

6.9.3 One Proof

I tend to provide proofs in the main text only if I feel that something can be learned about the underlying problem. For this book, while i do love the following proof, I am not sure it is valuable beyond seeing where the formulas come from.

Proof of OLS Estimators

This provides a proof that the estimators for the y-intercept and the slope (effect) are as stated in Section 6.5.1. The proof starts with the goal (minimizing the sum of squared errors) and uses calculus to achieve this goal.

Recall that the purpose of ordinary least squares regression is to obtain the estimators that minimize the squared errors. The word "minimize" suggests that we should take the derivative of the "sum of squared errors" with respect to the two parameters, set each equal to zero to find the critical values, then solve for the estimator. This is exactly what is done.

First, we start with determining the function we need to minimize and write it as a function of the two parameters (β_0 and β_1):

$$Q = \sum_{i=1}^{n} e_i^2$$
$$= \sum_{i=1}^{n} (y_i - \hat{y}_i)^2$$

Chapter Appendix

$$= \sum_{i=1}^{n} \left(y_i - (\beta_0 + \beta_1 x_i)\right)^2$$

$$= \sum_{i=1}^{n} (y_i - \beta_0 - \beta_1 x_i)^2$$

Next, we take the derivative of this function with respect to each parameter and set them equal to zero to obtain the critical values. First, the intercept:

$$\frac{\partial}{\partial \beta_0} Q = \frac{\partial}{\partial \beta_0} \sum_{i=1}^{n} (y_i - \beta_0 - \beta_1 x_i)^2$$

$$= \sum_{i=1}^{n} -2(y_i - \beta_0 - \beta_1 x_i)$$

$$0 \stackrel{set}{=} \sum_{i=1}^{n} -2(y_i - b_0 - b_1 x_i)$$

$$= \sum_{i=1}^{n} y_i - \sum_{i=1}^{n} b_0 - \sum_{i=1}^{n} b_1 x_i$$

$$= n\bar{y} - nb_0 - b_1 n\bar{x}$$

Thus, the OLS estimator of the intercept is

$$b_0 = \bar{y} - b_1 \bar{x}$$

Next, we take the derivative of Q with respect to β_1 to obtain the estimator of the slope:

$$\frac{\partial}{\partial \beta_1} Q = \frac{\partial}{\partial \beta_1} \sum_{i=1}^{n} (y_i - \beta_0 - \beta_1 x_i)^2$$

$$= \sum_{i=1}^{n} -2 x_i (y_i - \beta_0 - \beta_1 x_i)$$

$$0 \stackrel{set}{=} \sum_{i=1}^{n} -2 x_i (y_i - b_0 - b_1 x_i)$$

$$= \sum_{i=1}^{n} \left(x_i y_i - b_0 x_i - b_1 x_i^2\right)$$

$$= \sum_{i=1}^{n} x_i y_i - b_0 \sum_{i=1}^{n} x_i - b_1 \sum_{i=1}^{n} x_i^2$$

$$= \sum_{i=1}^{n} x_i y_i - (\bar{y} - b_1 \bar{x}) n\bar{x} - b_1 \sum_{i=1}^{n} x_i^2$$

$$= \sum_{i=1}^{n} x_i y_i - \bar{y} n \bar{x} + b_1 \bar{x} n \bar{x} - b_1 \sum_{i=1}^{n} x_i^2$$

$$b_1 \left(\sum_{i=1}^n x_i^2 - n\bar{x}^2 \right) = \sum_{i=1}^n x_i y_i - n\bar{x}\bar{y}$$

$$b_1 = \frac{\sum_{i=1}^n x_i y_i - n\bar{x}\bar{y}}{\sum_{i=1}^n x_i^2 - n\bar{x}^2}$$

Algebraically, this last line is equivalent to what we need to show. Thus, the two OLS estimators are

$$b_0 = \bar{y} - b_1 \bar{x}$$

$$b_1 = \frac{\sum_{i=1}^n (x_i - \bar{x}) y_i}{\sum_{i=1}^n (x_i - \bar{x})^2}$$

It may be interesting to for you to prove that this last formula is equivalent to the following

$$b_1 = \frac{\sum_{i=1}^n (x_i - \bar{x})(y_i - \bar{y})}{\sum_{i=1}^n (x_i - \bar{x})^2}$$

$$b_1 = \frac{\mathbb{C}ov[x,y]}{\mathbb{V}[X]} \qquad\qquad = \frac{\text{covariance}(x,y)}{\text{variance}(x)}$$

$$b_1 = \mathbb{C}or[x,y]\,\frac{s_y}{s_x} \qquad\qquad = \text{correlation(x,y)}\,\frac{\text{stdev}(y)}{\text{stdev}(x)}$$

The proofs that these are equivalent are just exercises in algebra.

6.9.4 Two Experiments in Election Statistics

When the mathematics gets rather difficult, it is helpful to fall back on experimentation to grow your understanding of the underlying statistics. This section looks at two such experiments. These are both designed to give a better understanding of coverage and how to estimate it from a problem.

Experiment 1: Coverage

The code for the first experiment. Located on page 125. Here, we are trying to determine which of the two methods for estimating the effective sample sizes is better. To do this, we estimate the coverage rate of each. Since we are claiming to calculate a 95% confidence interval in each case, we would expect the coverage rate to be rather close to 95%.

Chapter Appendix

```
##### Explicitly set the values
pi = 0.183
n1 = 1058; n2 = 1015; n3 = 1000; n4 = 1500; n5 = 2182

##### Perform the following 10,000 times, allowing us to get
#     10,000 estimates of the confidence interval endpoints,
#     allowing us to get a better understanding of the
#     differences in the two methods.

lwrBoundAvg = uprBoundAvg = numeric()
lwrBoundSum = uprBoundSum = numeric()

for(i in 1:10000) {

  # Generate one of each type of poll
  x1 = rbinom(1, size=n1, prob=pi)/n1
  x2 = rbinom(1, size=n2, prob=pi)/n2
  x3 = rbinom(1, size=n3, prob=pi)/n3
  x4 = rbinom(1, size=n4, prob=pi)/n4
  x5 = rbinom(1, size=n5, prob=pi)/n5

  # Calculate the two confidence intervals
  est = mean(x1,x2,x3,x4,x5) # The estimate

  avgSampleSize = mean(n1,n2,n3,n4,n5)
  sumSampleSize = sum(n1,n2,n3,n4,n5)

  lwrBoundAvg[i] = est-1.96*sqrt(est*(1-est)/avgSampleSize)
  uprBoundAvg[i] = est+1.96*sqrt(est*(1-est)/avgSampleSize)

  lwrBoundSum[i] = est-1.96*sqrt(est*(1-est)/sumSampleSize)
  uprBoundSum[i] = est+1.96*sqrt(est*(1-est)/sumSampleSize)

} ## End of the loop

##### At this point, the four confidence interval
#     boundaries will contain 10,000 values, one
#     for each of the generated sets of poll results.

### How many of these intervals actually contain pi?
sum( lwrBoundAvg<pi & pi<uprBoundAvg )/10000
sum( lwrBoundSum<pi & pi<uprBoundSum )/10000
```

These last two numbers are called the **coverage** rate of the procedure. It measures the proportion of the time that the confidence interval actually contains the true value of π. The rate should be close to the claimed coverage of 95%.

Here, we learn that the coverage rate for the averaging is 94%, which *is* close to our claimed value. The coverage rate for adding the sample sizes, on the other hand, is only 55%... *far from* our claimed value of 95%. Thus, we would conclude that averaging sample sizes is a better method than adding them.

Experiment 2: Coverage

The code for the second experiment. Located on page 127.

```
##### Explicitly set the values
pi = 0.183
n1 = 1058; n2 = 1015; n3 = 1000; n4 = 1500; n5 = 2182


##### Perform the following 10,000 times, allowing us to get
#     10,000 estimates of the confidence interval endpoints.

lwrBoundWtd = uprBoundWtd = numeric()

for(i in 1:1e4) {

  # Generate one of each type of poll
  x1 = rbinom(1, size=n1, prob=pi)
  x2 = rbinom(1, size=n2, prob=pi)
  x3 = rbinom(1, size=n3, prob=pi)
  x4 = rbinom(1, size=n4, prob=pi)
  x5 = rbinom(1, size=n5, prob=pi)

  # Calculate the confidence intervals
  totalFavoring = x1+x2+x3+x4+x5
  totalSample   = n1+n2+n3+n4+n5
  est = totalFavoring/totalSample
  lwrBoundWtd[i] = est - 1.96*sqrt(est*(1-est)/totalSample)
  uprBoundWtd[i] = est + 1.96*sqrt(est*(1-est)/totalSample)

} ## End of the loop

##### At this point, the four confidence interval
#     boundaries will contain 10,000 values, one
#     for each of the generated sets of poll results.

##### How many of these intervals actually contain pi?
sum( lwrBoundWtd<pi & pi<uprBoundWtd )/10000
```

7
Going Beyond the Two-Race

Throughout the book so far, we have focused on modeling a single candidate or position. Even when comparing two, we were able to reduce it to just one comparison. This is because multiple comparisons affect the coverage and confidence of multiple positions.

For instance, if I were to estimate the 95% confidence interval for Candidate A and the 95% confidence interval for Candidate B, then I should not directly compare the two confidence intervals to determine if Candidate A leads B or not. This chapter examines the underlying issue and offers two solutions to it. The first remains within the realm of Fisherian statistics but moves into three dimensions. The second uses Bayesian analysis.

Republic of The Gambia: The Presidential Election of 2016
 The Gambia is a small country in western Africa surrounding the lower reaches of the Gambia River (from whence it got its name). The Gambia gained its independence from the United Kingdom in 1965, but retained Queen Elizabeth as its monarch until 1970. The 1965 and 1970 republican referendums were hailed as examples of free and fair elections in Africa:

> ... but the results won widespread attention abroad as testimony to The Gambia's observance of secret balloting, honest elections, and civil rights and liberties.[Glo12, p28]

After that auspicious start, Gambian elections declined in quality—and frequency. Sir Dawda Jawara was elected president in 1970 and held office for over 24 years until he was deposed in a coup d'état. The resulting two years of military rule (led by Colonel Yahya Jammeh) gave way to a presidential election in 1996, which was unsurprisingly won by Colonel Yahya Jammeh, as was the 2001 election, as was the 2006 election, as was the 2011 election.

The run-up to the 2016 election looked a bit more promising to the opposition candidates, however.

The Gambia's Independent Electoral Commission (IEC), which is charged with overseeing elections and ensuring they meet the legal requirements, registered three candidates for the 2016 election: Jammeh, Adama Barrow (of the United Democratic Party), and Mama Kandeh (of the Gambia Democratic Congress). The remaining parties decided to unite behind Barrow to increase the probability that The Gambia would finally elect their third president.

December 1—election day—was peaceful and saw approximately 60% of the eligible voters cast their ballots for one of the three candidates, with the results illustrated in Figure 7.1.[Ind16]

7.1 Overview

Thus far, I have been careful to test elections with only two possible outcomes. The statistics is very clean and clear when estimating a single proportion—even when it appears one is estimating two. For instance, while the Scottish independence referendum had two positions they were complements of each other. You either voted *for* independence or you did not. Estimating the support for the status quo was as easy as estimating the support for independence and subtracting that from 100%.

Similarly, the Turkish 2023 general election (Chapter 5) focused only on the support for the Party of Greens and the Left Future (YSGP); the Bhutanese 2024 general election focused only on the People's Democratic Party (Chapter 3); the Korean 2017 election (Chapter 6) assiduously examined each candidate in turn, not comparing their position to the others; etc. I did this for a good reason: All of the statistical techniques covered are only usable when estimating a single population proportion, not comparing two of them from the same population.

In your introductory statistics course, you will have experienced procedures for understanding a single population proportion (see Section 3.1.2). You will also have experienced procedures for comparing two *independent* population proportions (Section 4.6.2). You will *not* have covered procedures for comparing two *dependent* proportions. This chapter looks at a few methods for doing just this.

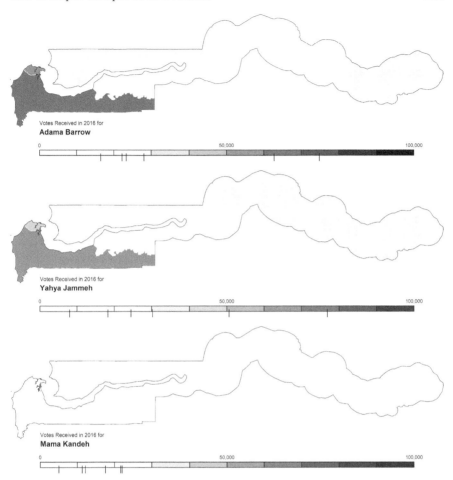

FIGURE 7.1
The outcome of the 2016 presidential election in Gambia. The maps are colored according to the *number* of votes cast for that candidate in the constituency. Source: The Gambia's Independent Electoral Commission.[Ind16]

7.2 The Multiple Comparisons Problem

Section B.2.9 explores the multiple comparisons problem at length. The following provides another way of seeing the problem.

For the sake of a running example, let us pretend that in November 22, 2016, TKS Polling surveyed 500 Gambian voters and asked each who they preferred in the upcoming election. Of those surveyed, 170 supported Barrow, 267 supported Jammeh, and the rest supported Kandeh. This leads to the following estimates of their support with a conservative margin of error of ±4.5 percentage points (pp):

- Barrow: 34.0%
- Jammeh: 53.4%
- Kandeh: 12.6%

These estimates are statistically sound. However, the usual next step—comparing the candidate support—is *not* statistically sound. We really *want* to compare the support level for Jammeh and Barrow and conclude that there is significant evidence that Jammeh is ahead.

We would typically do that by calculating the confidence intervals ($p \pm E$):

- Barrow: 29.5% to 38.5%
- Jammeh: 48.9% to 57.9%
- Kandeh: 8.1% to 17.1%

Then, since there is no overlap in the Barrow and Jammeh confidence intervals, we would conclude that there is a significant difference between the two candidate supports. However, you have just committed a cardinal sin in statistics: multiple comparisons without some adjustment. To see this, perform the following calculations assuming that the confidence intervals are correct:

- What is the confidence that the Barrow confidence interval contains his true support?

  (Ans: 0.95)

- What is the confidence that the Jammeh confidence interval contains his true support?

  (Ans: 0.95)

- What is the confidence that the Barrow confidence interval contains his true support *and* that the Jammeh confidence interval contains his true support?

  (Ans: $0.95 \times 0.95 = 0.9025$)

The Multiple Comparisons Problem

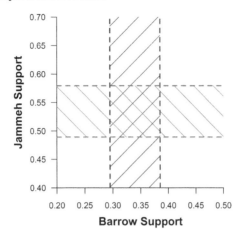

FIGURE 7.2
Illustration of the multiple comparisons problem. Note that the confidence widths of the two confidence intervals is 0.95, but the area of the doubly-shaded area is $0.9025 < 0.95$.

Figure 7.2 illustrates this problem.[1] The vertical axis are possible support values for Barrow; the horizontal, for Jammeh. The vertical and horizontal rectangular shaded areas represent the confidence intervals for Barrow and Jammeh support. These are defined as regions containing 95% of the probability. The central doubly shaded region is the intersection of these two confidence intervals. Its confidence area is $0.95 \times 0.95 = 0.9025$.

Thus, one fix is to widen the two confidence intervals so that the rectangle has probability 0.95. This observation leads to the Bonferroni Adjustment.

7.2.1 The Bonferroni Adjustment

A simple way to adjust the confidence level $(1 - \alpha)$ is to divide α by the number of things you are comparing. Thus, comparing two proportions, the confidence levels to use on the individual estimates is 0.975. This adjustment will ensure that the confidence level for the comparison is at least 95%.

So, to compare two candidate supports, change the confidence multiplier from 1.96 to 2.24, which changes the margin of error to ± 5.0pp. With this adjustment, the confidence intervals of Barrow and Jammeh overlap and we do not detect a difference in the two candidate supports.

---

[1] For those statisticians following along and checking up on me, you are 100% correct. There is actually more going on here. To make this more correct, one would need to show the bivariate Normal distribution. That would show the next level of difficulty: The two supports levels are *not* independent. Section 7.5 explores this a bit.

Dependence

Bonferroni is a general purpose adjustment that can be used in any case of multiple comparisons. Its strength is its flexibility. Its weakness is its power—or lack thereof. It can be shown that the Bonferroni adjustment will tend to create confidence intervals that are too wide. This has led to an entire subfield of statistics dedicated to improving upon Bonferroni.

Specifically, in this case, the intervals will be much wider than necessary. Section 7.5 discusses some effects of the correlation between voter support for two candidates.

7.3 The *Real* Problem

If you wish to delve deeper into the problem here (and my feelings wouldn't be hurt if you did *not*), then we need to draw something other than rectangles. The reality is that the joint distribution of support rate for these three candidates is a multivariate Normal distribution projected onto a two-dimensional regular simplex. If that statement doesn't scare you, continue reading.

As shown in Eqn. 3.1 (page 47), each support rate follows a Normal distribution. Together, the three follow a multivariate Normal distribution in three dimensions. The confidence interval for the three together becomes and ellipsoid in 3-space. The axes of the ellipsoid correspond to the standard deviations.[2]

However, there is a constraint on the support. The three must add to 1. In other words, if π_b is the support level for Barrow, π_j is the support level for Jammeh, and π_k is the support level for Kandeh, then we know

$$\pi_b + \pi_j + \pi_k = 1$$

Thus, this is just a projection onto the two-dimensional simplex—a triangle with vertices at $(0,0,1)$, $(0,1,0)$, and $(1,0,0)$. This observation may help us present current support estimates for the candidates in a meaningful manner to our readers.

Figure 7.3 is such an illustration. Note that the estimated support for the three candidates can be represented in a single graphic. The dark blob represents a 95% confidence region for candidate supports. The white dot indicates the sample proportions observed in this particular poll. That the

---

[2] If we use the conservative confidence intervals, the standard deviations are all the same, so the ellipsoid is really a sphere. So, that is one reason to use the conservative intervals.

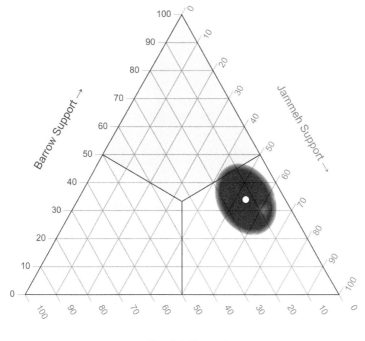

FIGURE 7.3
The estimated support level for the three Gambian presidential candidates with a 95% confidence ellipse. The confidence ellipse is calculated using the Wald method (Section 3.1.2). This graphic was created using the R packages PlotTools[Smi23], provenance[VRG16], and Ternary[Smi17].

white dot is in the "Jammeh" region of the graphic (bottom right) indicates that more of the people in the poll responded that they supported Jammeh than either of the other two candidates.

The confidence ellipse, on the other hand, includes some of the Barrow region (top). This indicates that we *cannot conclude* Jammeh is definitely ahead in the population (at the 95% confidence level). Note that The Gambia uses a plurality (first-past-the-post) electoral system to elect their president.

I have never seen a graphic like this used for illustrating polling results. One reason may be that it is more difficult to interpret than bar charts. Another reason is that this graphic does take up more space than do bar charts. In print media, space is costly.

With that being said, this graphic can *properly* answer some of the questions readers have about poll results. It can definitely illustrate it more correctly than bar charts. However, it does not answer *all* of those questions.

But, Bayes can.

7.4 The Bayesian Solution

We first encountered Bayesian analysis back in Chapter 3. Section 3.3 introduced **credible intervals** as a natural alternative to confidence intervals. Recall that credible intervals are directly tied to a clear probability statement. That is, we *can* state that the probability π is in a given interval is 95% (or 85% or 99% or what have you). That is, we do not have to "do the confidence dance" to properly interpret.

The previous two sections illustrate that there is a lot of complicated mathematics underlying the Fisherian paradigm. Furthermore, we are still dealing with indirectly determining who is in the lead and by how much. Once again, the Reverend Thomas Bayes rides to the rescue.

As said in Section 3.3.1, Bayesian analysis arises from Bayes' Law.[Bay63] If we let π be the population proportion and x be the data, then Bayes' Law can be written as

$$g(\pi \mid x) \propto \mathcal{L}(x \mid \pi) \, f(\pi)$$

They symbol $\propto$ means "is proportional to." When using this form, we need to ensure that the left side is a probability distribution. Frequently, we will do that simply by noticing that the product on the right "has the form of" a known distribution.

This led to the introduction of the Beta distribution as the conjugate prior for the Binomial likelihood. It made the math easy. The parameters of the posterior were just the parameters of the prior with the new data literally added. That is, if you write the prior distribution as $\pi \sim \text{BETA}(a, b)$ and if you observe x people supporting Candidate X and y people *not* supporting Candidate X, then the posterior distribution will be

$$\pi \mid x \sim \text{BETA}(a + x, \, b + y)$$

Framing it in this way suggests that there may be natural extensions to the Binomial and the Beta distributions. Luckily, there are: the Multinomial and the Dirichlet distributions. Each is the natural extension of the one-dimensional cases.

If you specify your prior information in the form of a Dirichlet and the data arise from a Multinomial distribution, then the posterior is also a Dirichlet. Specifically:

- Prior distribution: $\quad \pi \sim \text{DIR}(\alpha_1, \alpha_2, \ldots, \alpha_k)$
- Data likelihood: $\quad \mathbf{X} \sim \text{MULTINOM}(x_1, x_2, \ldots, x_k)$
- Posterior distribution: $\quad \pi \mid \mathbf{X} \sim \text{DIR}(\alpha_1 + x_1, \alpha_2 + x_2, \ldots, \alpha_k + x_k)$

The Bayesian Solution   159

I am not entirely sure that providing the probability functions for the Multinomial and Dirichlet distributions is helpful. If you think it is, then check out the relevant sections in Appendix B.

Ultimately, what is important is being able to get the computer to calculate quantities that you (and your audience) think are interesting. To see how to answer the "interesting" questions, let us have an example or two. As with the credible interval examples in Section 3.3, the hardest part is usually determining what needs to be calculated.

Example 1: Non-Informative Prior

Let us return to the 2016 presidential election in The Gambia. Recall that The Knox Poll had 170 Barrow supporters, 267 Jammeh supporters, and 63 Kandeh supporters. If we use a non-informative prior, then our prior is

- Prior distribution: $\pi \sim \mathrm{DIR}\,(1,1,1)$

The likelihood is

- Data likelihood: $\mathbf{X} \sim \mathrm{MULTINOM}\,(170, 267, 63)$

Thus, the posterior is

- Posterior distribution: $\pi \mid \mathbf{X} \sim \mathrm{DIR}\,(171, 268, 64)$

That was easy.

The "hard part" is that we are in the multidimensional realm. While multiple integration is an option, it is not the most generalizable. Thus, in order to approximate this posterior distribution, I will draw a very large sample from it and treat *that* as the population.[3]

```
prior1    = c(1,1,1)
obs       = c(170, 267, 64)
alpha1    = prior1 + obs
estimates1 = rdirichlet(1e6, alpha1)
```

After running this code, I have a sample consisting of a million (1×10^6, 1e6) elections from the $\mathrm{DIR}\,(171, 268, 64)$ distribution. Using a million should provide enough precision for our work. If your computer is strong enough, feel free to use 100 million (1e8). It will increase the precision to a fourth digit.

To see what is in the variable, or at least understand what is there, run the following line in R:

---

[3] Base R does not have a `rdirichlet` function available. However, there are many packages that provide it. For clarity, I use the one in the `gtools` package.[BWL22] Do not forget to run `library(gtools)` to activate the function.

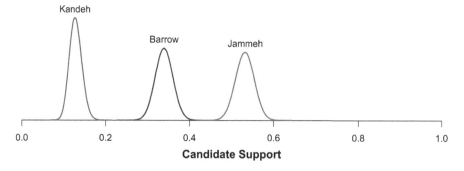

FIGURE 7.4
The estimated support level for the three Gambian presidential candidates based on their estimated posterior distributions based on the non-informative prior of Example 1.

```
head(estimates)
```

The head function lists the first six random draws from this distribution. This is what I got. Yours will differ because we are dealing with random sampling:

```
          [,1]      [,2]      [,3]
[1,] 0.3565256 0.5289708 0.1145036
[2,] 0.3134716 0.5515004 0.1350279
[3,] 0.3560479 0.5105110 0.1334411
[4,] 0.3360698 0.5252562 0.1386740
[5,] 0.3269515 0.5710759 0.1019725
[6,] 0.3039053 0.5580818 0.1380129
```

Being picky (and forgetful), I would like to label the three columns. This will help keep things straight should the analysis get complicated. To do this, I run

```
colnames(estimates) = c("Barrow","Jammeh","Kandeh")
```

Now, rerunning the head function shows that the columns are named. With that all done, I think we are ready to do some analysis. First, let us look at the distribution of supports (election outcomes) for the three candidates in Figure 7.4. This figure provides the posterior distributions for the support of each of the three candidates (Kandeh, Barrow, and Jammeh, respectively). There is no significant overlap between the three, and Jammeh's distribution is to the right of both, so one can conclude that Jammeh will win the election based on the data and the non-informative prior.

First, there are three probabilities I would like to calculate:

- What is the probability that Jammeh will receive at least 50% of the vote?

The Bayesian Solution

```
mean(estimates1[,"Jammeh"] >= 0.500)
```
93%

This probability is important for majoritarian electoral systems, one in which the winner is the candidate who receives at least 50% of the votes.

- What is the probability that Jammeh will receive more votes than Barrow?
```
mean(estimates1[,"Jammeh"] > estimates1[,"Barrow"])
```
100%

This probability is important for plurality electoral systems (a.k.a. first-past-the-post), one in which the winner is the candidate who receives the most votes, even if not a majority.

- What is the probability that Barrow will receive fewer than a third of the vote?
```
mean(estimates1[,"Barrow"] < 1/3)
```
38%

Alright, I'm not really sure what that last one means in terms of the Gambian election (the second is the important quantity). However, there are some electoral systems in which the candidate must receive a specific proportion of the votes to move on to the second round. In such a case, the third quantity is the probability that Barrow will not move on to the second round if there is a threshold of 33.3%.

There are electoral systems with even more complicated rules. For instance, in Bolivia, a presidential candidate wins the first round election if they *either* receive more than 50% of the vote *or* receive at least 40% of the vote and 10% more than the other candidates.[Cen24] If that were the electoral rule in The Gambia, we can calculate the probability that Jammeh avoids a run-off like this:

- What is the probability that Jammeh will avoid a run-off election if The Gambia has Bolivia's electoral system?
```
mean(estimates1[,"Jammeh"] > 0.50 |
  ( estimates1[,"Jammeh"] >= 0.40 &
    estimates1[,"Jammeh"] > estimates1[,"Barrow"] + 0.10
  )
)
```
93%

There are two new symbols in the above. The pipe | means "or." The ampersand & means "and." The parentheses and spaces make it easier for me to ensure that I am calculating the probability under the correct constraints.

Example 2: Informative Prior

Let us recalculate all of the probabilities using an informative prior. For this example, assume that I know something about the election and that my prior is

- Prior distribution: $\quad \pi \sim \text{DIR}(100, 40, 10)$

With this prior, and with rerunning most of the above code, the important probability is

- What is the probability that Jammeh will receive more votes than Barrow?
  ```
  mean(estimates2[,"Jammeh"] > estimates2[,"Barrow"])
  ```
  94%

Note that the probability did not change too much. This is because the data consists of 500 pieces of information, while this prior only has $100 + 40 + 10 = 150$ pieces.

Example 3: Highly Informative Prior

Now, to recalculate all of the probabilities using a highly informative prior. Let us assume that I *really* know something about the election and that my prior is

- Prior distribution: $\quad \pi \sim \text{DIR}(1000, 400, 100)$

With this prior, and with re-running most of the above code, the important probability is

- What is the probability that Jammeh will receive more votes than Barrow?
  ```
  mean(estimates3[,"Jammeh"] > estimates3[,"Barrow"])
  ```
  0.00%

Note that the probability changed a lot. This is because the data consists of 500 pieces of information, while the prior has three times as much: 1500 pieces. Thus, the estimates will reflect the prior more than the data.

The Bayesian Solution

Useful-*ish* Formulas

If $\pi \sim \text{DIR}(\alpha_1, \alpha_2, \ldots, \alpha_k)$, and the α_i add to n, then the expected value and variance for each π_i are

$$\mathbb{E}[\pi_i] = \mu = \frac{\alpha_i}{n}$$

$$\mathbb{V}[\pi_i] = \sigma^2 = \frac{\alpha_i(n - \alpha_i)}{n^2(n+1)}$$

Solving this system of two equations gives us a way to select α_i as a function of the expected value and standard deviation of π_i to represent the prior information:

$$\alpha_i = \frac{\mu^2}{\sigma^2}\left(\frac{1-\mu}{1+\mu}\right)$$

If you prefer to use the computer (as I do), this R function will calculate α for you given the expected value and standard deviation of π that you believe. Its code is given on page 168. Note that all standard deviations need to be the same.

```
dirichletPrior(mu, sigma)
```

Example 4: Another Informative Prior

Let us use the formula above to help us determine the proper prior parameters for the information I have.

Suppose I am rather certain that the support for Barrow is 49%; for Jammeh, 42%; and for Kandeh, 9%. Let us also suppose I am sure to a standard deviation of 2pp. That is, my belief for Barrow is from 43% to 47%, for Jammeh from 39% to 43%, and for Kandeh from 13% to 17%.

Using the `dirichletPrior` function for each candidate, I obtain a prior distribution of

- Prior distribution: $\pi \sim \text{DIR}(205.4547, 180.1268, 16.90596)$

With this prior, and with re-running most of the above code, the important probability is

- What is the probability that Jammeh will receive more votes than Barrow?
  ```
  mean(estimates4[,"Jammeh"] > estimates4[,"Barrow"])
  ```
  99.39%

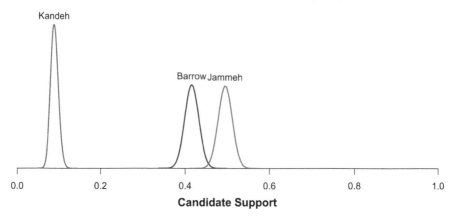

FIGURE 7.5
The estimated support level for the three Gambian presidential candidates based on their estimated posterior distributions based on the informative prior of Example 4.

Note that the probability changed just a bit from the non-informative prior. This is because the data consists of 500 pieces of information, while the prior only has 402 pieces.

Please compare the posterior distributions shown in Figure 7.4 (non-informative prior) to those in this example, Figure 7.5. The three curves have shifted a bit. Now, there is a little overlap between the Barrow and Jammeh curves. This suggests that the election may still be up for grabs. There is now only a 99.39% probability that Jammeh wins.

As mentioned *much* earlier (Section 3.3.1, page 60), the prior should contain all information known by the analyst—and no more. When done properly, a high-information prior mitigates the effects of a **poorly performed poll**. Large differences between the prior expected value and the observed sample proportion may actually reflect a bad poll... or a poorly chosen prior distribution.

7.5 A Brief Return to Fisher

Before we end this chapter, I need to return to a previous point and explain more fully. The footnote on page 155 states that Figure 7.2 overly simplifies the problem. That figure implies that the probabilities are independent, that

A Brief Return to Fisher

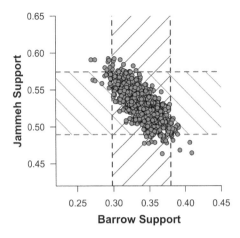

FIGURE 7.6
The estimated support level for the Barrow and Jammeh. Note that the two supports are not independent; there is a definite negative relationship between the two. The vertical dashed lines enclose 95% of the Barrow supports; the horizontal, 95% of the Jammeh supports. The central box contains 91.7% of the elections.

one merely needs to calculate areas to arrive at the conclusion. This is not true.

In Figure 7.2, the single-hatched regions both contain about 95% of the elections. But, for the central box to actually contain 90.25%, the two variables would need to be **independent**. This is clearly not the case as a vote for Barrow reduces the number of votes for Jammeh.

Thus, a better graphic illustration would be Figure 7.6. In that graphic, the single-hatched regions still both contain 95% of the election dots (reflecting that the intervals are both 95% intervals). However, the central box contains only 91.7% of the elections. So, if we are comparing the two candidate support levels, we would only be describing 92% instead of the claimed 95%. The Bonferroni adjustment would widen the two confidence intervals to ensure that the central box would contain at least 95% of the elections.

That adjustment would change the individual candidate confidence levels to 97.5% (= 1 − 0.05/2). The proportion of elections in the middle is now 95.7%, which is much closer to the claimed confidence level of 95%. That there are too many elections in the middle is a "feature" of the Bonferroni adjustment. It will tend to include too many, rather than too few.

Be aware that this discussion is important to the frequentists amongst us. Bayesians do not have to worry about these calculations and adjustments. In my estimation, this is the most important advantage to Bayesian analysis.

7.6 Conclusion

In previous chapters, we always estimated the support for a single candidate. Sometimes, we did this surreptitiously by estimating one position and comparing it to the complementary position. However, even in such cases, only one estimation was performed.

This is because the Fisherian (frequentist) statistics we all know and "love" only easily works with a single comparison. When doing multiple comparisons, one must know the distribution of all of the comparisons together. When that is computationally complicated, the general purpose Bonferroni adjustment is performed. This adjustment will always work. It will also tend to create confidence intervals that are too large.

The other difficulty with analyzing a race with more than two positions is graphically presenting the results of that analysis. Bar charts are a natural choice because they are easily understood and they are very good at showing the estimated support level for a single candidate or position. The trouble comes when people do what they shouldn't do: compare support levels between the bars.

This led to representing the support for each candidate (and its associated confidence interval) on a triangle. This allows the reader to compare support for all of the candidates at once. That strength, however, is tempered by two facts. The first is that this shape is rarely used to represent the current state of the race. Thus, the readers would have to learn how to interpret the graphic. The second is more daunting. The triangle only works for elections with only three positions. More positions would require shapes of higher dimension: four would require a tetrahedron (three dimensions); five, a "five cell" (four dimensions); etc. Representing four dimensions on a screen or a sheet of paper would be quite the challenge.

Bayesian analysis does not suffer from the multiple comparisons issue. At the end of performing Bayesian analysis, one had the entire distribution of all candidate supports. This allow one to directly calculate *probabilities* of specific outcomes, instead of Bonferroni-adjusted confidence intervals that only provide a set of reasonable values, by some definition of "reasonable."

This chapter marks the end of focusing on polls and what they tell us. The next couple of chapter look at what happens after the election is over. Can we test for evidence of unfairness? The next chapter introduces a method that is frequently **incorrectly** used to test for fraud in elections— the Benford Test. The chapter introduces it, discusses it, and ultimately dismisses it on the basis of logic.

Extensions 167

Chapter 9 introduces a direct test for one type of election unfairness: differential invalidation. In that chapter, a lot of time is spent tying the definition of "free and fair" to what it would look like if an election were not "free and fair." Once that is done, a test for differential invalidation is explored.

7.7 Extensions

In this section, I provide some practice work on the concepts and skills of this chapter. Some of these have answers, while others do not.

Review

1. Who won the 2016 presidential election in The Gambia? What does this say about the synthetic poll conducted by Knox College?
2. What are some strengths and weaknesses of illustrating the current support level for several candidates using a bar chart?
3. What is the "multiple comparisons problem"?
4. What is the Bonferroni adjustment? What is its purpose? What are its strengths and weaknesses?
5. What is a simplex? Why should on represent a race between three persons using a simplex?
6. What is a difference between the Binomial distribution and the Multinomial?
7. What is a difference between the Beta distribution and the Dirichlet?
8. How do we know that the support for Barrow and the support for Jammeh would be negatively correlated?

Conceptual Extensions

1. Using a die, devise a method to explain the multiple comparisons problem to others.
2. Find a country in the world with an interesting electoral system. Estimate the probability that Jammeh either wins the first round *or* passes through to the second round. The calculation should be similar to the lengthy one on page 161.

Computational Extensions

1. Let us assume that The Knox Poll surveys 1000 Gambians. In that sample, 547 claimed to support Barrow and 418 Jammeh. Calculate the probability that the Barrow wins the election using the following priors:

   (a) no prior information: $\alpha_b = \alpha_j = \alpha_k = 1$
   (b) moderately strong prior: $\alpha_b = 150$, $\alpha_j = 150$, and $\alpha_k = 150$
   (c) quite strong prior: $\alpha_b = 1500$, $\alpha_j = 1500$, and $\alpha_k = 1500$
   (d) strong prior: a standard deviation of 1%, and expected values of 45%, 40%, and 15%.

2. What do these different prior distributions tell us about the importance of being responsible and open when doing Bayesian analysis?

7.8 Chapter Appendix

This appendix contains some "review" material on random variables and their mean (expected value) and variance.

7.8.1 R Functions

No statistician does these calculations by hand. We all use the computer. It helps to increase our precision and reduce the time needed to do the analysis. Here are some functions that help with the analyses covered in this chapter.

- `colnames(x)`

  This function labels the column names for the table x. Usually, this is done for aesthetics and to make the analysis more clear. There is a sister-function that names the rows of a table, `rownames(x)`.

- `dirichletPrior(mu,sigma)`

  This function calculates the value of α_i for your prior distribution. Remember that there are k parameters that need to be calculated, so you will need to use this function for each α_i in your model. This is the code for this function:

  ```
  dirichletPrior = function(mu,sigma) {
      alpha = (mu^2 - mu^3)/(sigma^2 + mu*sigma^2)
      return( list(alpha=alpha) )
  }
  ```

Note that the standard deviation, `sigma`, needs to be the same for the separate priors.

- `head(x)`

  This shows the top six rows of a variable. There is a sister-function that shows the *bottom* six rows, `talk(x)`. In both cases, the number of rows shown can be specified like this: `head(x, 10)`.

- `mean(x)`

  Unsurprisingly, this returns the arithmetic mean of the given variable, x. However, if x is a series of TRUEs and FALSEs, then this returns the proportion that are TRUE.

8

Testing Elections: Digit Tests

This chapter begins looking at methods for detecting various types of electoral fraud and unfairness. Here, we look at what we can test if the government only provides vote counts for each of the candidates at the electoral division level. Some assert that this is sufficient information to test for fraudulent counting.

That is, the Benford test of this chapter allows us to gather evidence that the claimed vote counts were modified improperly. This test, and its generalization, have a rather limiting assumption underlying them. Ultimately, this makes them of questionable utility in election testing. The next chapter provides a better test.

Islamic State of Afghanistan: The Presidential Election of 2009
Afghanistan's second presidential election under its 2004 constitution took place on August 20, 2009. This election pitted incumbent Hamid Karzai against challenger Abdullah Abdullah of the United National Front and several minor candidates. The campaign saw the use of the state-run media and intimidation to affect the outcome. Public opinion polls gave Karzai a wide lead, but falling short of giving him the needed majority to avoid an October runoff election.

Election day was predictable. The Taliban called for a boycott of the "Western-led" poll. Violence broke out at several polling places. Challengers made charges of ballot-box stuffing and of false counting.[Gal09]

Two days later, the Independent Election Commission (IEC) announced its official results: Karzai received 54.6% of the valid votes cast (Figure 8.1).[Afg09b] The Electoral Complaints Commission (ECC), overseen by the United Nations and the final arbiter of the election, refused to validate the election until all fraud claims were adjudicated. However, even this did not progress smoothly. Methods of determining the proportion of votes to remove met with charges of Western meddling.

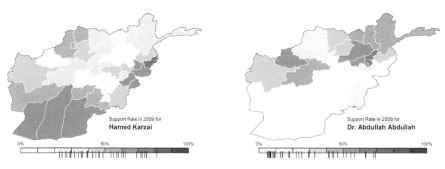

FIGURE 8.1
Map of the support for Hamid Karzai and Dr. Abdullah Abdullah in the 2009 Afghan presidential election. Note the complementary support regions. How is geography affecting these election returns? These data are from the former Afghan Independent Election Commission.[Ind09]

One thing was true: If there were a second round election, it would be between Karzai and Abdullah. Under US pressure, Karzai accepted a runoff election to be scheduled for November 7, 2009. Abdullah demanded certain members of the IEC be removed. They refused, Abdullah withdrew from the runoff election, and the IEC named Hamid Karzai President-elect.[Gal09]

Doubts over the extent of the fraud claims linger. However, because of the extant situation, the only numbers easily available are the vote counts at the province level (*wilāyat*). Are we able to use this information to test for unfairness in this election? If so, what types of unfairness?

Overview

Across the world, democratic elections are becoming the norm. Several non-democracies even hold elections to provide a veneer of popular legitimacy.[Hyd11, Wan99] As such, the presence of the word "democracy" does not necessarily indicate a democratic election. While Political Scientists do not agree on all of the specific requirements of a democratic election, they *do* agree that electoral fraud is not democratic.

While there are many ways to commit electoral fraud, two common ones are to stuff the ballot box and to change the vote counts. Ballot box stuffing takes place when the candidate's supporters place multiple completed ballots in the ballot box. This necessarily takes place at the precinct level.

Changing the vote counts can happen at any stage between the precinct and the central electoral commission, depending on the particulars of the electoral system. For instance, the ballots in the Afghan election of 2009 were air-lifted from the precincts to Kabul, where the electoral commission counted all of the ballots. Contrast this with Ghana, where the ballots are counted at each constituency collation center (with counts counter-signed by official party members) and only the totals are transmitted to the electoral commission in Accra.[Gha20]

Luckily, both of these fraudulent acts leave behind evidence. The trick is to find that evidence and determine if the evidence is due to fraud or merely to natural random variation in voting. This is where statistical analysis enters the picture.

8.1 Electoral Forensics

A goal of electoral forensics is to use statistical methods to determine if there is sufficient evidence to conclude that an election violates the democratic assumption— **the free and fair hypothesis**. Democratic elections must be free and fair. They must be free in the sense that voting is allowed. They must be fair in the sense that one person's ballot counts the same as any other person's ballot— or at least has the same *probability* of counting.

Unfortunately, the information needed to test for fraud is controlled by the very government being tested. This usually means the level of information available is quite limited. With this reality, electoral forensics seeks testing methods given the slight information available. Fortunately, many elections provide the vote counts for the candidates in each first-level administrative division. This is the data that Afghanistan's Independent Election Commission offers from the 2009 Presidential election.

Currently, the Benford test is the usual method for testing for violations of the free and fair hypothesis when only vote counts are available.[Meb10] The Benford test compares the actual digit frequencies to a hypothesized distribution, thus generating a p-value.

This is the strength of the Benford test. It offers a statistical test of the free and fair hypothesis based solely on the assumed distribution of vote counts. Its weakness is that there is **no reason** to believe that voting *should* be testable using the Benford test.

This chapter introduces the history of the original Benford test and its uses. It provides an update to the test, the generalized Benford test. It then explores its applicability to election data, concluding that it has severe shortcomings.

History and Insight 173

Ole's Warning

The inclusion of the Benford test in this book should not be construed as evidence that I support its use. In fact, as you read this chapter, you will see more and more that there are *significant* issues in applying it to elections.

To be completely clear: There is no reason to believe that the distribution of the initial digit of election returns has anything to do with the Benford distribution.

8.2 History and Insight

While browsing a book of logarithms (e.g., Figure 8.2), Harvard astronomer Simon Newcomb noticed something interesting:

> That the ten digits do not occur with equal frequency must be evident to any one making much use of logarithmic tables, and noticing how much faster the first pages wear out than the last ones. The first significant figure is oftener 1 than any other digit, and the frequency diminishes up to 9.[New81, p. 39]

From this observation, he derived the distribution of those leading digits. More importantly to him—and to us—he concluded that one could distinguish between a table of numbers "arising in nature" and their logarithms. The leading digits of the former follow his described distribution; of the latter, a uniform distribution.

Fifty-two years later, Frank Benford, a physicist and electrical engineer at General Electric, made the same observation:

> The pages containing the logarithms of the low numbers 1 and 2 are apt to be more stained and frayed by use than those of the higher numbers 8 and 9.[Ben38, p. 551]

Benford's contribution is not in this observation, but in its application:

> ...no one could be expected to be greatly interested in the condition of a table of logarithms, but the matter may be considered more worthy

0° 0′ 0″		0° 0′ 30″		60″ = 0° 1′ 0″	
Num.	Log.	Num.	Log.	Num.	Log.
0	−∞	30	·47712	60	·77815
1	·00000	31	·49136	61	·78533
2	·30103	32	·50515	62	·79239
3	·47712	33	·51851	63	·79934
4	·60206	34	·53148	64	·80618
5	·69897	35	·54407	65	·81291
6	·77815	36	·55630	66	·81954
7	·84510	37	·56820	67	·82607
8	·90309	38	·57978	68	·83251
9	·95424	39	·59106	69	·83885
10	·00000	40	·60206	70	·84510
11	·04139	41	·61278	71	·85126
12	·07918	42	·62325	72	·85733
13	·11394	43	·63347	73	·86332
14	·14613	44	·64345	74	·86923
15	·17609	45	·65321	75	·87506
16	·20412	46	·66276	76	·88081
17	·23045	47	·67210	77	·88649
18	·25527	48	·68124	78	·89209
19	·27875	49	·69020	79	·89763
20	·30103	50	·69897	80	·90309
21	·32222	51	·70757	81	·90849
22	·34242	52	·71600	82	·91381
23	·36173	53	·72428	83	·91908
24	·38021	54	·73239	84	·92428
25	·39794	55	·74036	85	·92942
26	·41497	56	·74819	86	·93450
27	·43136	57	·75587	87	·93952
28	·44716	58	·76343	88	·94448
29	·46240	59	·77085	89	·94939
30	·47712	60	·77815	90	·95424

FIGURE 8.2

A scan of a page from Richard Farley's *Tables of Logarithms*.[Far39] This is the first table in the book. Its level of wear can be inferred through the poor quality of the top edge of the book.

of study when we recall that the table is used in the building up of our scientific, engineering, and general factual literature. There may be, in the relative cleanliness of the pages of a logarithm table, data on how we think and how we react when dealing with things that can be described by means of numbers.[Ben38, p. 551]

In the course of his research, Benford gathered numbers from many sources and tested if the leading-digit distribution for each source followed his proposed distribution. These sources ranged from lengths of rivers to addresses in the telephone book to the mathematical sequence $\{n, n^2, n^3, \dots\}$. While the individual sources did not always follow the prescribed distribution (river lengths did, the mathematical sequence did not), Benford was able to find a commonality among those that did. Examining the table, Benford concluded that those sources of a random nature followed his "logarithmic law" much more closely than those sources arising from a mathematical formula:

> These facts lead to the conclusion that the logarithmic law applies particularly to those outlaw numbers that are without known relationship rather than to those that individually follow an orderly course; and therefore the logarithmic relation is essentially a Law of Anomalous Numbers.[Ben38, p. 557]

It is this distribution of leading digits of the integers from 1 to n that became the basis for using statistics to detect fraud and deviations from random ("natural") behavior.[Car88, CG07, Hil95, Ley96, Meb10, Nig11, Nig12]

8.3 The Benford Test

Many fraud tests, such as the Benford test, are based on the assumption that humans are *incorrectly* random. To illustrate this, I assigned one class to flip a coin 1000 times and record "H" if the coin came up "Heads" and "T" otherwise. Not surprisingly, many of them decided that writing down a series of "H" and "T" values was much easier and would give the "same results." However, as they discovered, it was straightforward to test if they actually flipped the coin.

This exercise (hopefully) led them to appreciate different "types of randomness" and how to compare one to another. While most realized that the number of heads flipped needed to be close to 500, they did not consider other aspects of the Binomial distribution. That no one in a class of 50 recorded *exactly* 500 heads was also a bit suspect.[1]

More importantly, they did not consider other aspects of the Binomial distribution.[2] For instance, if the number of heads follows a Binomial distribution, then we also know the distribution of the number of runs and of the run length.[3] That the students did not consider that such a distribution existed meant that fabricated data would most likely differ from the expected distribution.

This is *exactly* what happened.

---

[1]The probability of a single person getting exactly 500 heads is about 1 in 40. The probability of no one in a class of 50 getting exactly 500 heads is about 1 in 4.

[2]For a refresher on the Binomial distribution, please visit Appendix B.2 where the Binomial distribution is explored.

[3]A "run" is series of categorical outcomes that are the same. For instance, if a person flips the following sequence, then there are 3 runs: HHHTTHH. They have lengths 3, 2, and 2.

This is the idea underlying digit tests in electoral forensics. Election results, while not coin flips, should also demonstrate some divergence from the expected distribution if they are tainted by vote-count fraud (humans recording knowingly false vote counts).

How can vote-count fraud take place? While all electoral systems are unique in their specific structures, some generalities can be made. Frequently, the votes are not counted in a single, central location. Votes cast at the precinct level tend to be counted at the local level, with vote counts forwarded to some national electoral commission. There, the individual totals are added and the final announcements are made. Only later are the ballot papers shipped to the national electoral commission—if at all.

Ghana follows this strategy. When the polling station closes, the ballots are taken to a local facility, a Constituency Collaboration Center (CCC). The counting takes place at the CCC, in full view of candidate agents. When the ballots are counted, the totals are recorded in the "Blue Book" for that constituency, with the candidate agents signing off on the legitimacy of the count and receiving copies of the counts. Those vote counts are then forwarded to the Electoral Commission of Ghana in Accra, which collects them, totals them, and announces the official winners.[Gha12, §35]

Alternatively, Afghanistan did not follow this process in 2009; the polling places sent all ballot papers to a central counting facility in Kabul.[Afg09a, Article 12] In such cases, it is much easier to perpetrate vote-count fraud; one merely has to control the small group doing the centralized counting. In places like Ghana, the control must happen at the constituency level.

In countries like Afghanistan, changing the counts (or even creating false counts) is relatively easy. The person in charge of counting merely has to change the digits of one or more count. The digit most likely changed is the leading digit, as that digit has the most impact on the election. As mentioned above in my classroom, rarely are humans *correctly* random. They will attempt to force the appearance of randomness to hide their tracks, but they will use the wrong distribution. That is, the distribution of vote counts in fraudulent precincts is different than the distribution of vote counts in non-fraudulent districts.

The current digit test is the (integer) Benford test.[Hil95, Meb10]

Theorem 8.1. *The probability mass function of the Benford distribution, $BENF_1$, is*
$$\mathbb{P}\big[\,\mathcal{D}_1(X) = d\,\big] = \log\left(\frac{d+1}{d}\right)$$
where $\mathcal{D}_1(X) \in \{1, 2, \ldots, 9\}$ is the leading digit of the count.

For a proof of this, please see Lemma 8.7, page 195.

The Benford Test

TABLE 8.1
Vote counts for Hamid Karzai in several electoral districts for the 2009 presidential election in Afghanistan.

District	Karzai Count	Leading Digit
Badakhstan	74,656	7
Badghis	18,412	1
Baghlan	42,213	4
Balkh	109,776	1
Bamyan	45,788	4
Daikondi	38,879	3
Farah	25,200	2
⋮	⋮	⋮
Urozgan	14,395	1
Wardak	24,363	2
Zabul	7,327	7

The Benford distribution is used in forensic accounting to detect fraud in expense accounts [Car88, CG07, Nig11, Nig12] and in elections to detect vote-counting fraud.[Meb10] Theoretically, the distribution of the leading digit of votes for a given candidate follows the integer Benford distribution. As humans tend to randomize much more uniformly, vote-count fraud will produce a distribution different from the Benford distribution. As such, a simple chi-squared test can be (and is) used to detect this type of election fraud.[Car88]

8.3.1 Using the Benford Test

Before looking at the appendix to this chapter to see the proof (if you are so inclined), read through the following example. In performing the Benford test, there are three steps: determine the leading digits of the vote counts, create a frequency distribution of those leading digits, and use the Chi-square test to compare the observed distribution to the hypothesized one (the Benford distribution).

Before we perform these three steps, I note that it is often helpful to see the data being tested. To that end, Table 8.1 partially provides data from the 2009 Afghan election, which can be found in the `afg2009pres.csv` data file. The leading digits are determined in the table, and their frequency distribution is shown in Figure 8.3. In fact, that graphic illustrates how much the observed frequency distribution of the leading digits differs from what is expected—if Benford's Law holds perfectly. The → corresponds to the hypothesized (Benford) distribution, the ← corresponds to the observed distribution. The vertical segments illustrate the differences between the two.

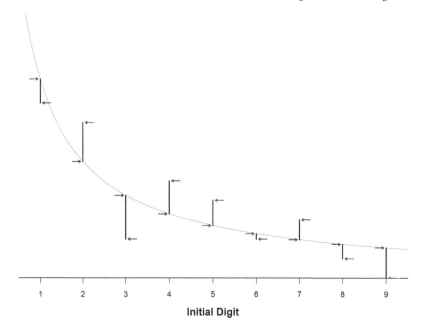

FIGURE 8.3
The observed initial digit distribution as compared to the expected under the Benford distribution. The → corresponds to the hypothesized (Benford) distribution, the ← corresponds to the observed distribution.

A usual statistical test for comparing an observed discrete distribution with a hypothesized distribution is the Chi-square goodness-of-fit test.[4] It compares the scaled squared deviances to the Chi-square distribution of $g-1$ degrees of freedom (number of groups minus 1, where $g=9$ as we are dividing the leading digits into 9 groups, one per digit).

Applying the test to this data results in a p-value of 0.6503. Since this number is greater than our usual value of alpha, $\alpha = 0.05$, we cannot claim that this test found evidence of electoral fraud. In other words, this result does not provide enough evidence to assert that the deviation from the Benford distribution is due to anything other than random chance.[5]

I leave it as an exercise to calculate the p-values for the other three major candidates. Note that in this data file, Ahmadzai is "`Ashraf.Ghani.Ahmadzai`," Abdullah is "`Dr..Abdullah.Abdullah`," Karzai is "`Hamed.Karzai`," and Bashardost is "`Ramazan.Bashardost`."

---

[4]While this is arguably the usual statistical test, there are others that may work better for smaller sample sizes.[NP33, RC88, WW97]
[5]The appendix to this chapter (Section 8.8.4 on page 199) provides the R code for this analysis.

The Benford Test

TABLE 8.2
The calculated p-values for the Benford test on the four main candidates in the 2009 Afghan Presidential election. That none of these are less than $\alpha = 0.05$ indicates that this test did not detect any fraud.

Candidate	p-value
Ashraf Ghani Ahmadzai	0.8841
Dr. Abdullah Abdullah	0.7269
Hamed Karzai	0.6503
Ramazan Bashardost	0.3618

Table 8.2 provides the results I obtained.

Note that none of the calculated p-values is sufficiently small to enable us to claim evidence of fraud. In other words, all differences between the observed distribution of leading digits and the Benford distribution are within expected randomness. Thus, there is no evidence of electoral fraud here, either.

Statistics is Evidence

This raises an interesting question: What if, after testing all of the candidates, none of the p-values are lower than α? Can we conclude there was no fraud?

No. We do not have sufficient *evidence* that there was fraud. This is not surprising. Statistics is all about amassing evidence—both for and against—on claims made about reality. The amount of evidence generated depends on a few factors. One is the amount of data available. Another is the quality of the data. A third is the quality of the *test*. It is this third that leads statisticians to developing more and more tests.

Again, allow me to emphasize that there are multiple reasons a statistical test may not detect fraud. Only one of which is that there really is no fraud. The other two reasons are based on weaknesses arising from probability and statistics: the sample size is too small and/or the test is too weak.

8.4 Extending Newcomb and Benford

In the previous section, we looked at the Benford distribution as stated by Newcomb and Benford. In both cases, the researcher assumed the upper bound was not relevant to the distribution. This happens when the upper bound in (election turnout) in *each division* is either an integral power of 10 or is infinite. Neither of these two cases is likely in any realistic election. This section looks at using the Generalized Benford Test, which makes neither assumption.

8.4.1 The Generalized Benford Distribution

The case of Lemma 8.7 (page 195) was the familiar result from Newcomb and Benford.[Ben38, Meb10, New81] Both Newcomb and Benford allude to this second (more general) case, but neither one significantly explored it nor derived it.

Theorem 8.2. *The probability mass function of*

$$BENF_1(0, \theta) := \mathbb{P}\big[\, \mathcal{D}_1(X) = d \mid \theta \,\big]$$

is

$$\frac{\lfloor \theta \rfloor}{\theta} \cdot \log\left(\frac{d+1}{d}\right) + \begin{cases} 0 & \lfloor 10^{\theta - \lfloor \theta \rfloor} \rfloor < d \\ \frac{1}{\theta}\left(\theta - \lfloor \theta \rfloor + \log\left(\frac{1}{d}\right)\right) & d \leq \lfloor 10^{\theta - \lfloor \theta \rfloor} \rfloor < d+1 \\ \frac{1}{\theta}\log\left(\frac{d+1}{d}\right) & d+1 \leq \lfloor 10^{\theta - \lfloor \theta \rfloor} \rfloor \end{cases}$$

and is continuous in θ, where $\theta := \log X$.

Because the proof does not give much insight into the test, I am throwing it in the chapter appendix (page 197). I encourage you to read through it if you want to see how the description of the probability function is translated into a mathematical statement about it.

By the way, since the symbol will be used in the future, $\lfloor \theta \rfloor$ is the "floor" of θ, It is the greatest integer that less than or equal to θ.

> **Aside.** *If $\theta \in \mathbb{N}$, Theorem 8.2 reduces to Lemma 8.7, as it should. To see this, note that $\theta = \lfloor \theta \rfloor$ when $\theta \in \mathbb{N}$ and that $10^0 \leq d$ for all $d \in \{1, 2, \ldots, 9\}$.*

Extending Newcomb and Benford

Both Newcomb and Benford proved their results in the case where the values of X were unbounded; i.e. when $\theta \to \infty$. Using Theorem 8.2, we can now prove their assertions differently.

Corollary 8.3. *As the upper bound gets larger and larger, $\theta \to \infty$, the generalized Benford distribution converges in distribution to the integer Benford distribution; that is,*
$$BENF_1(0, \theta) \xrightarrow{d} BENF_1$$

I will prove this corollary here. It does give some insight into what is happening, as does Figure 8.4. Furthermore, this proof allows me the opportunity to show the importance of the Theorem of Carabinieri.[6]

I encourage you to work through the proof while looking at the figure.

Proof. From Theorem 8.2, we can see that $\mathbb{P}\big[\,\mathcal{D}_1(X) = d \mid \theta\,\big]$ varies between a lower bound at $d = 10^{\theta - \lfloor \theta \rfloor}$ and an upper bound at $d + 1 = 10^{\theta - \lfloor \theta \rfloor}$.

For the lower bound, substitution gives:

$$\text{Lower bound} = \frac{\lfloor \theta \rfloor}{\theta} \log\left(\frac{d+1}{d}\right) + \left(1 - \frac{\lfloor \theta \rfloor}{\theta} - \frac{\theta - \lfloor \theta \rfloor}{\theta}\right)$$

$$= \frac{\lfloor \theta \rfloor}{\theta} \log\left(\frac{d+1}{d}\right)$$

Thus
$$\text{Lower bound} \to \log\left(\frac{d+1}{d}\right), \text{ as } \theta \to \infty.$$

This last step arises because $\lfloor \theta \rfloor$ is bounded between θ and $\theta - 1$. Thus, $\frac{\lfloor \theta \rfloor}{\theta}$ is bounded between $\frac{\theta}{\theta} = 1$ and $\frac{\theta - 1}{\theta} = 1 - \frac{1}{\theta}$. As $\theta \to \infty$, both bounds go to 1.

I leave the proof of the upper bound as an exercise, because it follows the same steps as for the lower bound.

Finally, since the upper bound converges to the lower bound, the "Two Policemen and a Drunk" Theorem allows us to conclude that
$$BENF_1(0, \theta) \xrightarrow{d} BENF_1(0, n)$$

---

[6] This is a very popular and useful theorem. Evidence suggests that it was first developed (discovered?) by Eudoxus of Cnidus, with Archimedes of Syracuse developing it further.

As with many ancient Greek mathematical results, this theorem was rediscovered and expanded by Carl Friedrich Gauss in the 19th Century. Modern mathematicians know it as the "Squeeze" Theorem or as the "Two Policemen and a Drunk" theorem, depending on how difficult college was for them.

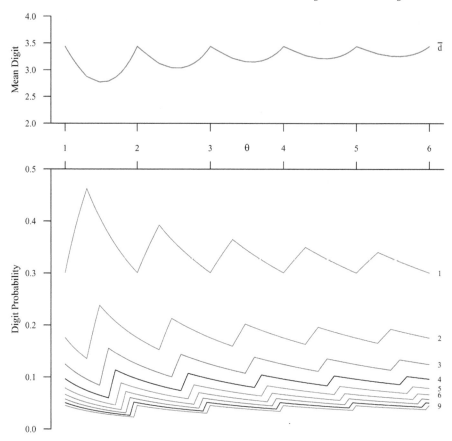

FIGURE 8.4
Means and probabilities for various values of θ. The top plot is the mean leading digit; the bottom, leading digit probabilities with probabilities for the digit 1 at top, and the rest in decreasing order.

for $n \in \mathbb{N}$, as $\theta \to \infty$.

∎

Reviewing Figure 8.4 shows some mathematical results. Recall that it provides a graphical display of the leading digit probabilities for values of θ through $\theta = 6$ (turnouts from 1 to 1 million voters). The top graph demonstrates how the mean leading digit varies with the value of θ. Note that $\mathbb{E}[\,D\,] \to 3.440$ as $\theta \to \infty$.

Two Important Take-Aways

The importance of this observation is two-fold. First, the Benford test makes the assumption that the leading digit distribution is constant. Figure 8.4 shows that it varies, and *does so* by a significant amount. Not having the right expected distribution means that all statistical tests based on it are wrong.

The second importance is in the fact that the *means* vary (are a function of θ). The Benford distribution states that the mean is a constant 3.440. However, Figure 8.4 (top) shows that this is not the case. For instance, if the turnout (upper bound) is 3500, the actual mean is only 3.153, and the variance is 5.657. These differences shift and narrow the intervals, making those calculated under the Benford distribution inaccurate.

And so, using the generalized Benford distribution is much more appropriate than using the usual Benford distribution. It is based on the same assumptions as the Benford, but without the requirement that the upper bound is an integer power of 10.

8.5 Using the Generalized Benford Distribution

Now that we have a sufficient understanding about the generalized Benford distribution, it is time to turn to using it in electoral forensics. This section explores three ways of implementing the new distribution, as well as why none of the three are perfect.

8.5.1 The Problem

The Benford test assumes that the digit probabilities do not depend on the turnout (Section 8.3). This makes testing easy because it is straight-forward to determine the expected distribution of the initial digits; it is the same for all turnouts. The *generalized* Benford distribution, on the other hand, has different digit probabilities for each turnout. Since different electoral divisions have different turnouts in the election, this makes determining the proper distribution of the leading digits in the entire election rather problematic. In other words, each election has n distributions to combine in some manner, one for each electoral division.

Three methods come to mind to solve this problem: simulating the likelihood and two manners of averaging the digit distributions across the divisions. The following sections explore each of these options.

8.5.2 The Likelihood Simulation Method

One method to unify these n tests, albeit computationally intensive, is to create a test statistic and estimate its distribution through simulation. This is a typical method in statistics when the distribution of the statistic is unknown or too complicated.

To this end, let us define θ_i such that 10^{θ_i} is the total number of votes cast in electoral division i. Let V_i be the vote count in electoral division i for a specific candidate, and let $d_i = \mathcal{D}_1(V_i)$ be the leading digit of the vote count in division i for that candidate. If the leading digit follows the generalized Benford distribution, then the probability mass function in division i is given by Theorem 8.2, with θ indexed by the electoral division, i.

The likelihood of observing these data (the vote counts in all of the individual divisions) is the product of the individual probabilities:

$$\mathcal{L}(\theta; \mathbf{d}) = \prod_{i=1}^{n} \left(\mathbb{P}[\, \mathcal{D}_1(V_i) = d_i \mid \theta_i\,] \right)$$

Here, the symbol $\prod$ indicates that we should multiply the things to the right across the index i. This is very similar to the use of the $\sum$ symbol, which tells us to *add* the things to the right.

Because these likelihoods will be quite small (very close to zero), it is better that we work with the logarithm of the likelihoods, the "log-likelihoods." That way, instead of dealing with numbers on the order of 10^{-50}, for example, we will be dealing with numbers on the order of -50. While computers may round the former to 0, they will not significantly round the latter.

The log-likelihood is calculated as

$$l(\theta; \mathbf{d}) = \sum_{i=1}^{n} \log \left[\mathbb{P}[\, \mathcal{D}_1(V_i) = d_i \mid \theta_i\,] \right]$$

The choice of base for the logarithm is not important, as long as there is consistency. Because much of this discipline uses the common logarithm, I will use base 10.

The Process

To estimate the distribution of the log-likelihood, one will do the following many times.

1. For each division in the election:
   (a) generate a leading digit in that division according to the total number of votes cast in that division

(b) calculate the probability of obtaining that leading digit according to the generalized Benford distribution
(c) calculate the logarithm of that probability
2. Finally, add all of the log-probabilities for that particular pseudo-election

Each trip through this algorithm produces a log-likelihood value for the real election. Doing this process many, many, many times will provide a simulated distribution for the election. Now that you have an estimated distribution of the log-likelihood for that particular election, you can calculate a 95% confidence interval to determine the amount of evidence in favor of the election being free and fair.

The next examples illustrate this procedure.

Example: Oklahoma 2008

As a first example, let us turn to the 2008 US Presidential election in Oklahoma.[Wi20a] In the data file ok2008pres.csv, the vote counts are aggregated at the county level ($n = 77$). The observed log-likelihood is -59.75168 for McCain, the winner of Oklahoma. Using simulation with 1,000,000 replications, the approximate distribution of the log-likelihoods is given in Figure 8.5, Left Panel.[7] From this distribution, we can estimate a 95% confidence interval to be from -70 to -59. Thus, because the observed values for *both* John McCain (M on the figure axis) and Barack Obama (O) are in this interval, there is no evidence of vote count fraud in the 2008 US Presidential election in Oklahoma.

This is the code I used to generate these elections.[8]

```
loglikelihood = numeric()
B = 1e6

for(elxn in 1:B) {
  logprob=0
  for( division in 1:divisions) {
    dgb  = dGBenford(turnout[division])
    vote = sample(1:9, size=1, prob=dgb)
    logprob = logprob + log10(dgb[vote])
  }
  loglikelihood[elxn] = logprob
}
```

---

[7]Note that simulating a million elections does take quite a while on a typical computer (over 3 hours on my laptop). The time requirement is more reasonable to simply simulate 10,000. The resulting loss of precision is relatively minor. However, since the final conclusions are serious, that extra precision is worth the time.

[8]The code for the function dGBenford can be found in the code appendix for this chapter, page 192. It simply calculates the probability density for a given threshold according to the generalized Benford distribution.

This is the code I used to estimate the 95% confidence interval.

```
|| quantile(loglikelihood, c(0.025, 0.975) )
```

Code Comments. The first line sets aside memory for the variable that will contain the one-million election likelihoods.[9] The second specifies that the number of iterations through the algorithm is a million (1×10^6).

The next block of code is divided into two parts. The inner part goes through each electoral division, determines the digit distribution according to the generalized Benford distribution, generates a random leading digit for that division, determines the probability of observing that digit, and adds its log to the election probability variable, `logprob`.

The outer block goes through this process for the one million simulated elections (B = 1e6). Thus, at the end, the variable `loglikelihood` will contain a million values, each being the logarithm of the probability of observing that particular election outcome (series of leading digits).

The `quantile` function determines the endpoints of that central 95% of simulated observations. Thus, this line tells us that the 95% confidence interval is all log-likelihoods between -70 and -59.

Figure 8.5 (left) shows the simulated distribution of the log-likelihoods for this election. At the base, the observed log-likelihood for Obama (O) and McCain (M) are provided.

Example: Afghanistan 2009

As a second example, let us examine the 2009 Afghan Presidential election.[Ind09] Recall that these votes are aggregated at the *wilāyat* level, $n = 34$. The observed log-likelihood is -28.93255 for Karzai (K on the figure) and -30.87798 for Dr. Abdullah (A). Simulating a million elections provides an estimated distribution of the likelihoods (Figure 8.5, Right Panel) leading to a 95% confidence interval from -32 to -25. Thus, this test offers no evidence of vote count fraud in this election.

Example: Lithuania 2009

In 2009, Dalia Grybauskaitė stood for reelection in Lithuania. She was heavily favored and reportedly received 69.1% of the votes cast. The votes are aggregated at the level of the municipality ($n = 60$).[Vyr09] Her observed log-likelihood is -50.4279 ("G" on Figure 8.6, Left Panel). Again, simulating a million elections provides an estimated distribution of the likelihoods leading

---

[9]Technically, the variable will hold the logarithm of those likelihoods, base 10.

Using the Generalized Benford Distribution 187

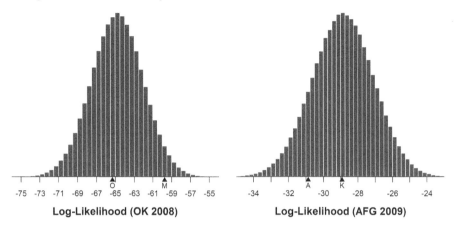

FIGURE 8.5
Estimated distributions of the log-likelihood for the Oklahoma 2008 election (left panel) and for the Afghan 2009 election (right panel). The observed log-likelihoods for the individual candidates are provided. "O" and "M" represent Obama and McCain in the left panel; "A" and "K" represent Abdullah and Karzai in the right.

to a 95% confidence interval from -55 to -46. Thus, this test also offers no evidence of vote count fraud in this election.

Note that Grybauskaitė was heavily favored in the election (her closest rival received 58% fewer votes). There were also no claims of fraud. Thus, this particular conclusion is not surprising. Since its independence from the Soviet Union in 1990, Lithuania has quickly become a liberal democracy.[Bel03] Free-and-fair elections are their norm.

Example: Côte d'Ivoire 2010

As a final example of the likelihood simulation method, let us examine the 2010 Ivorian Presidential election. Recall that this was the first presidential election held in Côte d'Ivoire since the end of the 2002–2007 civil war. While the first round had over a dozen candidates, none received a majority of the votes cast. This required a second round to ensure one candidate received a majority. November's run-off had President Laurent Gbagbo face off against Alassane Ouattara.

As expected in the aftermath of a civil war, the election results were highly contested. The Constitutional Council concluded that Gbagbo won. However, the Independent Electoral Commission concluded that Ouattara won. Because of the differences, the Second Ivorian Civil War began. After the death

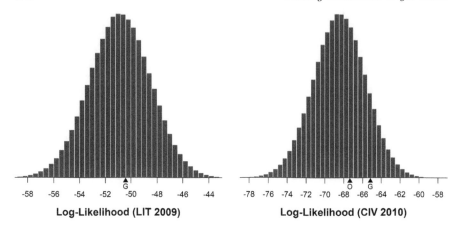

FIGURE 8.6
Estimated distributions of the log-likelihood for the Lithuanian 2009 election (left panel) and for the Ivoirian 2010 election (right panel). The observed log-likelihoods for the individual candidates are provided. "G" repesents Grybauskaitė in the left panel; "O" and "G" represent Ouattara and Gbagbo in the right.

of thousands, forces loyal to Ouattara, including many in the international community, captured Gbagbo and brought this chapter to a close.

According to this analysis, the observed log-likelihoods are -65.18116 (Gbagbo) and -67.35902 (Ouattara).[Com10] The 95% interval for the log-likelihoods is from -74 to -63. From this, there is no evidence that the results posted by the Independent Electoral Commission are fraudulent.

8.5.3 Multinomial Averaging

In lieu of using simulation to estimate the distribution of a test statistic, we can estimate the test statistic in another way. Recall that the leading digit in each electoral division has a different distribution, which is a function of θ_i. That these θ_i are typically unique makes creating the "correct" distribution quite difficult. However, two simple methods present themselves for approximating that "correct" distribution, for combining the n distributions into one.

1. Average the θ_i and use this value in the generalized Benford distribution.
2. Average the n separate generalized Benford distributions.

Using the Generalized Benford Distribution

Both options have the benefit of being rather fast to perform; there is no need to simulate elections. Both also benefit from using more aspects of the data (expected leading digit frequencies) than does the likelihood simulation method (only an average). Thus, both of these methods should be more powerful than the previous.

Four Examples The following illustrate the two averaging methods, with the first showing how to code it in R. I encourage you to modify the code appropriately to obtain the results for the other elections.

The first block strips off the only information needed from the data, the counts for John McCain and the total number of votes cast in the election. The second block parses the data to create the observed leading digit frequency table (obs1 and obs2). It also calculates the expected distribution according to the two multinomial averaging methods described here. The first averages the θ_i, while the second averages the n generalized Benford distributions in the election. Finally, the last block performs the Pearson chi-square test.[10]

```
jmc = ok$MCCAIN
tot = ok$TOTAL

obs  = getInitDigitDistribution(jmc)
exp1 = findMultinomialProbabilities1(tot)
exp2 = findMultinomialProbabilities2(tot)

chisq.test(obs, p=exp1)
chisq.test(obs, p=exp2)
```

Table 8.3 provides the p-values (last two columns) for the main candidates in the four elections. As usual in frequentist statistics, the p-value should be compared to a pre-selected value of α, the acceptable Type I error rate.[11] The default choice is $\alpha = 0.05$. However, I recommend a smaller one, perhaps even $\alpha = 0.001$. Claiming an election is fraudulent is serious business.

Thus, these two tests both conclude that there is no significant evidence of counting fraud in these elections at the $\alpha = 0.001$ level. These tests agree with the likelihood simulation method.

Again, remember the admonition from page 179: Statistics is about gathering evidence. The quality of the evidence depends on the quality of the data and of the test. The data are only as good as the government. However, since we are using it to *test a government claim*, using their data against them is

---

[10] Note that the chi-square test assumes the expected number of counts in each digit to be at least 5. That means the results are very approximate when the number of electoral divisions is fewer than $n = 100$. This is the cause of the warning message:
`Chi - squared   approximation   may   be   incorrect`

[11] In general, a Type I error occurs when one rejects a true null hypothesis. In this context, however, a Type I error occurs when the election is fair, but the test concludes it is not. In other words, one claims the government is cheating, but they are not.

TABLE 8.3
The results from performing both types of multinomial averaging. The p-values for the hypothesis that the leading digits follow the generalized Benford distribution are provided in the last two columns.

Election		Candidate	MA I	MA II
Oklahoma 2008	[Wi20a]	McCain	0.4444	0.4769
		Obama	0.1814	0.1486
Afghanistan 2009	[Ind09]	Abdullah	0.1936	0.2023
		Karzai	0.5874	0.5266
Lithuania 2009	[Vyr09]	Grybauskaitė	0.0034	0.0030
Côte d'Ivoire 2010	[Com10]	Gbagbo	0.3824	0.3045
		Ouattara	0.6876	0.5471

not a problem. The tests, however, are little-explored in terms of the Type I error rate. Furthermore, Forsberg suggests that all three of these tests are of relatively low power (high Type II error rate).[For14]

Specious at Best

More tellingly, Forsberg also concludes that tests based on the Benford and/or generalized Benford distribution lack even face validity. After all, why *should* the leading digit of votes counts follow this specific distribution? Until this last question is answered, claims of count fraud based on Newcomb or Benford or Forsberg are specious *at best*.

8.6 Conclusion

One type of fraud perpetrated is for the official in charge of counting to falsify the records. While this is most easily done in systems where the raw ballots are transported to a single central agency before counting, it can be done if a large number of precinct officials are corrupt. Detecting this type of fraud relies only on the vote counts and the turnout in each electoral division. Frequently, this is the only information provided by the government.

To test for vote-count fraud, one must assume the true distribution of an election. It is here that Newcomb (1881) and Benford (1938) made significant contributions to fraud detection with the Benford distribution. Unfortunately,

Extensions 191

the Benford distribution makes the assumption that the electoral division sizes are exact powers of 10 (or infinite), which is not the case. To fix this, Forsberg (2014) formulated the generalized Benford distribution.

While the generalized Benford distribution may work easily for testing the leading digits of counts from electoral divisions of equal size, it is not directly applicable for electoral divisions of different sizes in the same way that the Benford distribution is. The remainder of the chapter explored methods for using the generalized Benford distribution to test for vote-count fraud.

In the next chapter, we move on from digit tests to regression tests. In *this* chapter, we only had vote counts to analyze. In the next, we include an additional variable: invalidation rates. Under the free and fair hypothesis, these two variables should be independent. Thus, Chapter 9 will examine how to test this assumption in the presence of electoral fraud.

8.7 Extensions

In this section, I provide some practice work on the concepts and skills of this chapter. As usual, some of these have answers, while others do not. Before diving into these, I do suggest reading the appendix. There is some code that that will make the calculations much easier.

Review

1. Why would it be easier for counting fraud to take place in the 2009 Afghan election than a typical Ghanaian election?
2. What conditions are necessary for an election to be democratic?
3. How did Newcomb discover the Benford distribution?
4. What is the difference between "natural" and "unnatural" numbers according to Benford?
5. What is the purpose of the Chi-square test?
6. Why does ignoring information in the data tend to lead to a less-powerful test?

Conceptual Extensions

1. Flip a fair coin 1000 times and create a histogram of run length. Estimate the average run length. Now, simulate those 1000 flips by

hand. That is, randomly write out H and T values as you *think* would arise from flipping a coin. Create a histogram of run length and determine the average run length in your fake data. Compare the results.

2. Why is the expected distribution of initial digits not uniform? What is the intuition behind the Benford distribution and test?
3. Why is 10 the usual base of the logarithm used in the Benford test? Would using a different base work equally well?
4. What is the "Two Policeman and a Drunk" theorem and why is it useful in proofs?
5. In Figure 8.4, at what points is the mean digit the greatest?
6. Why would one expect the likelihood simulation method of Section 8.5.2 to produce a less-powerful test than either of the two multinomial averaging methods of Section 8.5.3?
7. Which of the two multinomial averaging methods of Section 8.5.3 would you expect to be more powerful?
8. What conclusions can be drawn from Table 8.3?

Computational Extensions

1. Find the populations of all the countries in the world. Write down the initial digit of those populations. Determine if that initial digit follows the Benford distribution sufficiently closely.
2. Find the populations of $n = 100$ cities in the world. Write down the initial digit of those populations. Determine if that initial digit follows the Benford distribution sufficiently closely.

8.8 Chapter Appendix

This section contains the R functions, the mathematical proofs, and the excess code used in this chapter. While the code may not seem important to learn, it does help you work through the logic of the calculations.

Chapter Appendix 193

8.8.1 R Functions

This chapter introduced some functions ultimately designed to make your calculations easier. As I tell my students, let the computer do the calculations; that is what it is designed to do.

To help, here is a listing of the functions and their meanings. More information about these functions can be found in the online help for each.

- `chisq.test(o, e)` This performs the usual Pearson's chi-squared test. This test compares the observed counts (o) to the hypothesized probability distribution (e) to determine if it is reasonable to assert that the observations came from that distribution. Please read Section 8.8.4 for its use.

- `dGBenford(x)` This calculates the digit probabilities for a given population size. Please read Example 8.5.2 for its use.

- `findMultinomialProbabilities1(x)` This calculates the *expected* probabilities for each digit. It averages the district populations to obtain a single value of θ. With that value, it calculates the probabilities using the generalized Benford distribution. Please read Section 8.5.3 for its use.

- `findMultinomialProbabilities2(x)` This calculates the *expected* probabilities for each digit. Using the generalized Benford distribution, it calculates the digit probabilities for each district, then returns the average across districts for each digit. Please read Section 8.5.3 for its use.

- `getIntegerDigitDistribution(x)` This calculates the observed frequency distribution of the leading digits in the counts provided. It is frequently used as a first step in testing elections using the Benford test. Please read Section 8.8.4 for its use.

- `read.csv(path)` This function reads a comma-separated values (csv) file located at the path you provide. Since the function just reads the file, you should save the results as a variable (with a meaningful name).

- `sample(digits, size, prob)` This provides a sample of size *size* taken from the digits *digits* without replacement. The probabilities for selecting each digit is given in the *prob* variable. Please read Example 8.5.2 for its use.

8.8.2 Derivation of the Benford distribution

The chapter started out with a description of the Benford's origin, then jumped to an example. For those interested in the proof, here it is.

To determine the probability mass function of the Benford distribution, we return to Newcomb's observation that the earlier pages of logarithm books were more worn than later pages; that is, those pages dealing with numbers beginning with a 1 were used more often. From this, Newcomb deduced that the probability of a given digit being the leading digit of a number was $\log(d+1) - \log(d)$. To get to this point, it is necessary to define a distribution and a function.

Definition 8.4 (Log-Uniform Distribution). *Define Y as a Uniformly distributed random variable from 0 to θ, $Y \sim \text{UNIF}(0, \theta)$. Then $X := 10^Y$ has a log-Uniform distribution, symbolized as $X \sim \text{Log-UNIF}(0, \theta)$, with sample space $X \in [1, 10^\theta]$.*

That is, the log of a log-uniform distribution follows a Uniform distribution, hence the name.

This distribution has a cumulative distribution function $F_X(x) = \frac{\log x}{\theta}$ and probability density function $f_X(x) = \frac{\log e}{x\theta}$, where "ln" and "log" are the natural (base e) and common (base 10) logarithm functions, respectively. Figure 8.7 provides the probability density function for the log-Uniform distribution over its support (for $\theta = 6$, a turnout of 1 million people).

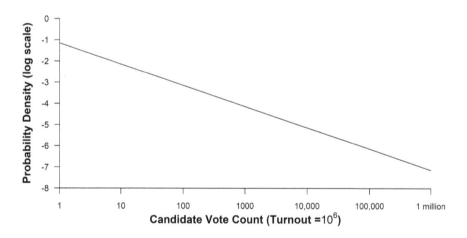

FIGURE 8.7
The probability density function of the Log-UNIF(0, 6) distribution over its support, $[1, 10^6]$. Note that the vertical axis is in log-units. Thus, -4 corresponds to 0.0001.

Chapter Appendix 195

Definition 8.5 (Leading-digit Function). *Define the leading-digit function $\mathcal{D}_1(x)$ taking a non-negative real number as its argument and returning the first (leading) digit of that number:*

$$\mathcal{D}_1(x) := \begin{cases} \lfloor x \, 10^{-\lfloor \log x \rfloor} \rfloor & x > 0 \\ 0 & x = 0 \end{cases}$$

Here, $\lfloor \cdot \rfloor$ is the "floor" (a.k.a, the "greatest integer") function. It rounds down. For example, $\lfloor 1.3 \rfloor = 1$, $\lfloor 4 \rfloor = 4$, and $\lfloor -4.2 \rfloor = -5$.

Finally, with the log-Uniform distribution (Defn. 8.4) and the leading-digit function (Defn. 8.5) we can define the Benford distribution:

Definition 8.6 (Benford Distribution). *If $X \sim$ Log-UNIF$(0, \theta)$, then the distribution of $\mathcal{D}_1(X)$ is the Benford distribution (of order 1), written as*

$$D \sim BENF_1(0, \theta)$$

If $\theta \in \mathbb{N}$, then we can drop the parameters of the Benford distribution and just specify $D \sim \text{BENF}_1$, which is termed the "integer Benford distribution."

Finally, we can determine the probability mass function of the Benford distribution. First, note that there are cases. In the first, $\theta \in \mathbb{N}$. In the second, $\theta \notin \mathbb{N}$. The derivation in the first case is simple; in the second, a bit more complicated.

Possibility 1: $\theta \in \mathbb{N}$

Let us now deal with the first possibility. Here, $\theta \in \mathbb{N} := \{1, 2, 3, \dots\}$. This is the simple case as the probability function of the Benford distribution turns out to be independent of θ.

Newcomb [New81] provided a proof of this based on a limiting equi-spaced circular distribution. Benford's proof [Ben38] relied on a counting argument. Both are worth the read if you are interested in the mathematics.

The following proof relies on the cumulative distribution function.

Lemma 8.7. *The probability mass function of the integer Benford distribution, $BENF_1(0, \theta) := \mathbb{P}\big[\, \mathcal{D}_1(X) = d \mid \theta \,\big]$, is independent of θ and is $\log\left(\frac{d+1}{d}\right)$ for $\theta \in \mathbb{N}$. This distribution is also abbreviated as $BENF_1$*

Proof. Recall $X \sim$ Log-UNIF$(0, \theta)$ and $F_X(x) = \frac{\log x}{\theta}$. Then,

$$\mathbb{P}\big[\, \mathcal{D}_1(X) = d \mid \theta \,\big]$$

is equivalent to calculating the area under the pdf curve corresponding to values with leading digit d. Equivalently, this is summing up over integer powers of 10 the differences in the CDF values of $(d+1) \times 10^i$ and $d \times 10^i$. That is:

$$\mathbb{P}[\mathcal{D}_1(X) = d \mid \theta] = \sum_{i=0}^{\theta-1} \left(F_X\left((d+1) \times 10^i\right) - F_X\left(d \times 10^i\right) \right)$$

$$= \sum_{i=0}^{\theta-1} \left(\frac{\log\left((d+1) \times 10^i\right)}{\theta} - \frac{\log\left((d+1) \times 10^i\right)}{\theta} \right)$$

$$= \sum_{i=0}^{\theta-1} \left(\frac{\log(d+1) + i}{\theta} - \frac{\log d + i}{\theta} \right)$$

$$= \sum_{i=0}^{\theta-1} \left(\frac{\log(d+1)}{\theta} - \frac{\log d}{\theta} \right)$$

$$= \frac{1}{\theta} \sum_{i=0}^{\theta-1} \left(\log(d+1) - \log d \right)$$

$$= \frac{1}{\theta} \, \theta \left(\log(d+1) - \log d \right)$$

$$= \log(d+1) - \log d$$

$$= \log\left(\frac{d+1}{d}\right)$$

And this is what we were to prove.

■

The probability mass function (pmf) of the integer Benford distribution, BENF$_1$, is graphed in Figure 8.8. Note the differences in probabilities among the digits. In fact, there is a *six-fold* difference in probability between getting a "1" and a "9".[12]

Additionally, note that the probability distribution does not depend on the actual value of θ, as long as it is a positive integer. This makes the Benford test easy to implement. The generalized Benford test (Section 8.4) suffers from being much more complicated to use.

---

[12] If you expected the digits to be much more uniform in their likelihood, you are not alone. Also, you would have randomized incorrectly, allowing us to catch your fraud!

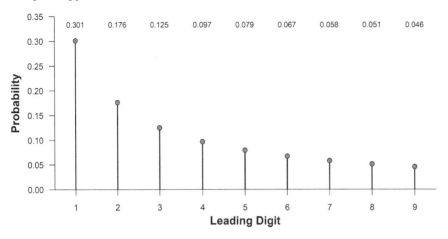

FIGURE 8.8
Distribution of lead digits according to the integer Benford distribution, $BENF_1$. Note that the lead digit is not uniformly distributed.

8.8.3 Generalized Benford Proof

This section of the appendix proves the generalized Benford test.

Theorem 8.1. *The probability mass function of the generalized Benford distribution*

$$BENF_1(0,\ \theta) := \mathbb{P}\big[\ \mathcal{D}_1(X) = d \mid \theta\ \big]$$

is

$$\frac{\lfloor \theta \rfloor}{\theta} \cdot \log\left(\frac{d+1}{d}\right) + \begin{cases} 0 & \lfloor 10^{\theta - \lfloor \theta \rfloor} \rfloor < d \\ \frac{1}{\theta}\left(\theta - \lfloor \theta \rfloor + \log\left(\frac{1}{d}\right)\right) & d \leq \lfloor 10^{\theta - \lfloor \theta \rfloor} \rfloor < d+1 \\ \frac{1}{\theta} \log\left(\frac{d+1}{d}\right) & d+1 \leq \lfloor 10^{\theta - \lfloor \theta \rfloor} \rfloor \end{cases}$$

and is continuous in θ, where $\theta = \log X$.

To prove this, note that there are three possible cases, each depending on the digit under consideration, d, and the quantity $\lfloor 10^{\theta - \lfloor \theta \rfloor} \rfloor$. This latter quantity is the leading digit of the turnout. The first possible case happens when the observed digit is less than d. The second is when the observed digit *is* d. The final is when the observed digit is greater than d.

Proof. With respect to the function, there are three cases. In the first case, the leading digit of X, the turnout, is less than the digit under consideration. In the second case, the leading digit of X is the digit under consideration. In the third case, the leading digit of X is greater than the digit under consideration.

This proof uses the definition of the cumulative distribution function (CDF) and algebra:

Case 1: In this case, the leading digit of X is less than the digit under consideration.

$$\mathbb{P}\big[\,\mathcal{D}_1(X) = d \mid \theta\,\big] = \sum_{i=0}^{\lfloor \theta \rfloor - 1} F\big((d+1) \times 10^i\big) - F\big(d \times 10^i\big)$$

$$= \lfloor \theta \rfloor \left(\frac{1}{\theta}\big(\lfloor \theta \rfloor + \log(d+1)\big) - \frac{1}{\theta}\big(\lfloor \theta \rfloor + \log d\big) \right)$$

$$= \frac{\lfloor \theta \rfloor}{\theta} \left(\big(\lfloor \theta \rfloor + \log(d+1)\big) - \big(\lfloor \theta \rfloor + \log d\big) \right)$$

$$= \frac{\lfloor \theta \rfloor}{\theta} \big(\log(d+1) - \log d \big)$$

Thus,

$$\mathbb{P}\big[\,\mathcal{D}_1(X) = d \mid \theta\,\big] = \frac{\lfloor \theta \rfloor}{\theta} \log\left(\frac{d+1}{d}\right)$$

Case 2: In this case, the leading digit of X is the digit under consideration.

$$\mathbb{P}\big[\,\mathcal{D}_1(X) = d \mid \theta\,\big] = \sum_{i=0}^{\lfloor \theta \rfloor - 1} F\big((d+1) \times 10^i\big) - F\big(d \times 10^i\big)$$
$$+ F\big(10^\theta\big) - F\big(d \times 10^{\lfloor \theta \rfloor}\big)$$

$$= \frac{\lfloor \theta \rfloor}{\theta} \log\left(\frac{d+1}{d}\right) + F\big(10^\theta\big) - F\big(d \times 10^{\lfloor \theta \rfloor}\big)$$

$$= \frac{\lfloor \theta \rfloor}{\theta} \log\left(\frac{d+1}{d}\right) + \left(\frac{\log\big(10^\theta\big)}{\theta}\right) - \left(\frac{\log\big(d \times 10^{\lfloor \theta \rfloor}\big)}{\theta}\right)$$

$$= \frac{\lfloor \theta \rfloor}{\theta} \log\left(\frac{d+1}{d}\right) + \frac{\log 10^\theta}{\theta} - \frac{\log d + \log 10^{\lfloor \theta \rfloor}}{\theta}$$

$$= \frac{\lfloor \theta \rfloor}{\theta} \log\left(\frac{d+1}{d}\right) + \frac{\theta}{\theta} - \frac{\log d + \lfloor \theta \rfloor}{\theta}$$

$$= \frac{\lfloor \theta \rfloor}{\theta} \log\left(\frac{d+1}{d}\right) + 1 - \frac{\log d}{\theta} - \frac{\lfloor \theta \rfloor}{\theta}$$

Chapter Appendix

Case 3: In this case, the leading digit of X is *greater than* the digit under consideration.

$$\mathbb{P}[\,D_1(X) = d \mid \theta\,] = \sum_{i=0}^{\lfloor\theta\rfloor-1} F\Big((d+1) \times 10^i\Big) - F\Big(d \times 10^i\Big)$$
$$+ F\Big((d+1) \times 10^{\lfloor\theta\rfloor}\Big) - F\Big(d \times 10^{\lfloor\theta\rfloor}\Big)$$
$$= \frac{\lfloor\theta\rfloor}{\theta} \log\left(\frac{d+1}{d}\right) + F\Big((d+1) \times 10^{\lfloor\theta\rfloor}\Big) - F\Big(d \times 10^{\lfloor\theta\rfloor}\Big)$$
$$= \frac{\lfloor\theta\rfloor}{\theta} \log\left(\frac{d+1}{d}\right) + \frac{\log(d+1) + \lfloor\theta\rfloor}{\theta} - \frac{\log d + \lfloor\theta\rfloor}{\theta}$$
$$= \frac{\lfloor\theta\rfloor}{\theta} \log\left(\frac{d+1}{d}\right) + \frac{1}{\theta} \log\left(\frac{d+1}{d}\right)$$

The theorem's formula follows from algebra. Continuity follows from the function values at the end points being equal.

∎

Note that the probability function itself is easily stated. However, since the probabilities also depend on the number of people voting in the district, using this distribution for testing is rather problematic, as Sections 8.5.2 and 8.5.3 show.

8.8.4 Miscellaneous R Code

The following are the R code for several of the analyses covered in the chapter.

Afghanistan, 2009

This the R code to perform the usual Benford test for the 2009 Afghan presidential election:

```
# Load data
afg = read.csv("afg2009pres.csv")

# Know the variables
names(afg)

# Access just the Karzai vote counts
candidate = afg$Hamed.Karzai

# Determine the distribution of Karzai's leading digit
obsDist = getInitDigitDistribution(candidate)
```

```
# Perform the Chi-square test
chisq.test( obsDist, p=log10(1+1/(1:9)) )
```

In the chi-squared test, the first slot goes to the observed frequency distribution of vote counts. The second slot (p=) goes to the hypothesized distribution. The result of this test is

```
        Chi-squared test for given probabilities

data:  obsDist
X-squared = 5.9727, df = 8, p-value = 0.6503
```

To test the other candidates, just change the "`candidate`" line to name the specific candidate code as revealed in the "`names(afg)`" line.

9

Testing Elections: Regression Tests

There are a couple of other methods that can be used to cheat in an election, both are examined in this chapter. One method is differential invalidation; the other, ballot box stuffing. Both methods leave evidence that can be detected using regression.

This chapter looks at regression methods that can test for both. The first is typically taught in introductory statistics courses, so it may be familiar. The other three are more advanced in their mathematics, if not their application here.

Republic of Côte d'Ivoire: The Presidential Election of 2010
After almost a decade of civil war and five years of postponed elections, the people of Côte d'Ivoire went to the polls in 2010 to elect their fifth president since independence from France in 1960. Three major candidates presented themselves for the first round: Henri Bédié, the republic's second president who ascended to the position after the death of long-time leader Félix Houphouët-Boigny in 1995; Alassane Ouattara, a former prime minister who left his position in 1993 at the urging of Bédié; and incumbent President Laurent Gbagbo, whose presidency spanned the civil war that embroiled the country for eight years.

The first round of the election was held on October 31. According to their electoral system, if no candidate obtained a majority of the votes cast, there would be a runoff election between the two leading vote recipients. As expected, things did not proceed smoothly: militias and the remnants of the northern rebellion made voting in the west and the north rather dangerous; and candidates accused each other of fraud. To add to the confusion, different "official" sources reported different vote totals for these three candidates. The official source, the Independent Electoral Commission (CEI), gave Bédié third place with only 25.2% of the vote. More importantly, it concluded that neither Gbagbo nor Ouattara had received a majority of the votes cast, meaning that a second round of voting would need to be held.[YÌ0]

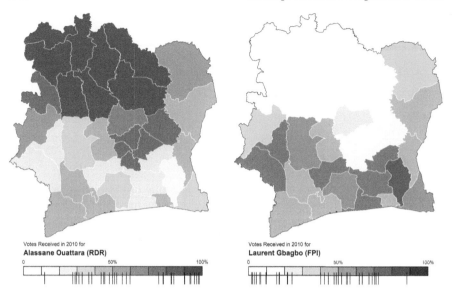

FIGURE 9.1
Maps of the support for Ouattara and for Gbagbo. Note the regionalism apparent in the support maps above. These data are from The Ivoirian *Commission Electorale Independante*.[Com10]

The November 28 second-round election featured President Gbagbo and Prime Minister Ouattara. Ouattara accused Gbagbo of causing the civil war, while Gbagbo claimed Ouattara had planned two coup attempts. Supporters of each candidate were largely divided along tribal/geographic lines, as Figure 9.1 suggests.

The CEI began to announce tallies as they became available. However, the country's divisions found their way even into the allegedly independent CEI, with one member snatching official vote counts from another during a press conference. No results were announced that night.

Again, each side accused the other of vote fraud. Security forces sought to stifle the violence, but merely added to it. By the time the CEI announced that Ouattara won the election with 54% of the vote, the two sides were in open conflict. To make matters worse, the Constitutional Council declared the CEI had no legitimacy and announced President Gbagbo had won the election. Although the international community recognized Ouattara as the legitimate president, it took foreign intervention and six more months of civil war before Gbagbo left the presidency, allowing Ouattara to take the helm of the deeply divided country.[BBC11a]

Differential Invalidation 203

And yet, the question remains. Two official Ivoirian agencies reported different election outcomes. The Constitutional Council claimed the CEI returns were fraudulent. To rectify this, the Constitutional Council invalidated all votes in the seven northern provinces—Ouattara's stronghold.[Age10] Did the Constitutional Court have evidence of electoral fraud? If so, what was it?

Overview

In the previous chapter, we explored a few tests of the reported vote count. Since that was the only information available, those tests compared the observed count to a hypothesized distribution. In this chapter, we will look at a test that can be done if the government provides the candidate vote counts *and* the number of invalidated ballots. That is, we will examine a set of tests comparing the invalidation rate to the candidate support rate. Under the free and fair hypothesis, these two variables should be independent. If not, then the invalidation rate depends on for whom the ballot was cast. A violation of the free and fair hypothesis may be due to an unfairness in the electoral system or to electoral fraud.

These tests are all variations of regression tests. As such, this chapter begins with a review of regression schemes, examining ordinary least squares (OLS) regression and its assumptions. From there, three other methods are briefly examined, each being slightly better than the last. Those three are weighted least squares (WLS), Binomial regression through generalized linear modeling (GLM), and Betabinomial regression through vector generalized linear modeling (VGLM).

9.1 Differential Invalidation

Before jumping into the mathematics of regression, it is very important to define differential invalidation, what it means, and why regression would be an appropriate test.

In elections, voters cast ballots. Counting authorities count those ballots. This counting may take place at the precinct level (as in Ghana, [Gha12, §35]) or at some centralized location (as in Afghanistan, [Afg09a, Article 12]). While counting ballots, a counting authority may decide to declare an individual ballot invalid—uncountable. There are two main reasons to invalidate a ballot. The first is that the ballot is actually improperly cast or marked. The second

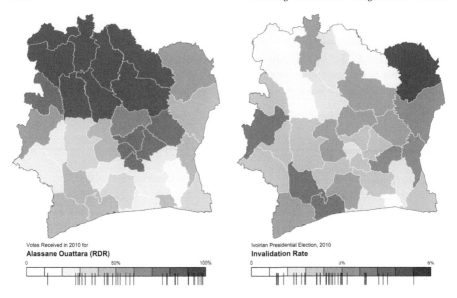

FIGURE 9.2
Maps of the support for Ouattara and the invalidation rates. Note that there seems to be a relationship between the two variables. Those regions with high support for the challenger tended to have their ballots invlaidated at a higher rate. These data are from The Ivoirian *Commission Electorale Independante*.[Com10]

is that the ballot is cast *for* the "wrong" person or *by* the "wrong" person. The former is legal, while the latter is differential invalidation.

Definition 9.1 (Differential Invalidation). *The act of invalidating a ballot based on demographic attributes or on who the ballot was cast for.*

Do the two maps of Figure 9.2 suggest differential invalidation based on whom the ballot was cast for? That is, do we see a relationship between invalidation rate and candidate support rate?

Quick Example

In a small division, there were 30 votes for Candidate A and 35 votes for Candidate B. I, a strong supporter of Candidate A, decide to invalidate 6 of Candidate B's votes. My actions give my preferred candidate the win, 30 to 29. Yay!

This is an example of differential invalidation in that I invalidated ballots based on whom the ballot was cast for.

Differential Invalidation

If we had the actual ballots *and* watched the counter invalidate the ballots *and* asked the counter why the ballot was invalidated *and* the counter said because it was cast for the wrong person, *then* we would have incontrovertible proof of differential invalidation. However, we cannot rely on the dishonest counter being honest. We can only work with the data provided by the government, which tends to include only the candidate votes, the number of invalidated ballots, and the total ballots cast aggregated at the electoral division level.

This summary data makes detection rather difficult, as we must rely on large sample sizes and hope that the differential invalidation is both large-scale and systemic. Differential invalidation taking place in just a few divisions will be nigh impossible to detect using statistical methods.

However, if the differential invalidation is wide-spread and severe, we should be able to detect its consequences. How? The following not-so-quick example starts us on our way to a method.

9.1.1 Toward a Test

Let our small electoral division have a referendum. Of the 70 votes cast, 28 were in favor of the referendum, 35 were against it, and 7 ballots were invalidated because the marks were smudged and voter intent could not be discerned (a legitimate reason). A table of the real, unreported results is given here.

Vote	Count	Percent	Percent
Yes	28	40.0%	44.4%
No	35	50.0%	55.6%
Invalidated	7	10.0%	

Here, the first percent is "percent of all ballots cast," and the second percent is "percent of all valid (counted) ballots cast." Note that the results would be reported as "55.6% to 44.4% against the referendum."[1]

Now, let us suppose I am a ballot counter who is staunchly in favor of the referendum. Because I want to, I decide to invalidate a further 10 "No" ballots. I am sure I could find some reason to do this. Maybe the voters for these 10 ballots only checked the box next to the "No" position and the law wants it fully filled in. Perhaps there were stray markings on those 10 ballot papers and I decided I couldn't determine the will of the voters. Perhaps I just decided to throw them out for no reason other than they were cast for the wrong position. Whatever the reason, I invalidated those 10 ballots using

---

[1] A consequence of this reporting is that it would be assumed that those ballots invalidated had the same distribution of referendum support as those counted. If this assumption is not correct, then what can we determine about the ballots? By the way, this is not a rhetorical question. It provides some insight into differential invalidation.

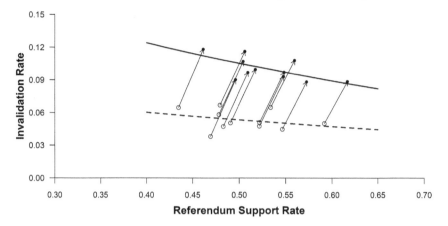

FIGURE 9.3
Change in the position of the electoral divisions in the example. The circles mark the "real" positions, while the solid discs represent the positions after differential invalidation took place. The curves are the regression curves summarizing the relationship between the two variables.

different "rules of invalidation" than I used for those ballots for the "Yes" position.

The following table presents the official results under this scenario.

Vote	Count	Percent	Percent
Yes	28	40.0%	52.8%
No	25	35.7%	47.2%
Invalidated	17	24.3%	

Note the effect of my differential invalidation of the reported results. First, when the percentages were calculated out of all ballots cast, the percentage for the position I was against declined and the invalidation rate increased. There was no change in the percentage for the position I favored, "Yes." However, when the position-support percentages were calculated out of only the number of valid votes, all three changed.

Figure 9.3 illustrates this with 11 electoral divisions. The circles correspond to the real vote results; the solid discs, to the reported results after differential invalidation took place. The arrows show the effect of differential invalidation in that electoral division. Note that the ultimate effect is that the regression curve also changes in shape. When the election lacked systemic differential invalidation (open circles), the regression curve was rather flat

Differential Invalidation

(dashed), showing no significant relationship between the two variables. However, when differential invalidation took place (closed circles), the regression curve showed a significant relationship (solid line).

Thus, our test for systemic differential invalidation should rely on some sort of regression test, looking for that relationship. If the regression curve is essentially flat, then there is no evidence of differential invalidation. If, however, it is sloped, then there is evidence of differential invalidation.

The next sections explore our regression options, along with the strengths and weaknesses of each. The first is the one typically introduced in Introductory Statistics. The others tend to be covered in higher-level regression courses, if at all.

9.1.2 Another Example

Before we get started on the mathematics (or get started on skimming over the mathematics), let's see a quick example of testing for differential invalidation. In the previous example, we saw that differential invalidation leads to patterns in the "**invalidation plot**."[2] In this section, let us see how to leverage this observation.

First, let's start with a free and fair election. This will ultimately allow us to discover the effects of differential invalidation (and of ballot box stuffing) on an election. Here, let our election have 10 electoral divisions of varying turnout and varying support for the "Support" position and varying levels of invalidation:

```
registered = c(544,579,648,650,721,756,761,794,853,978,1033)
turnout    = c(256,185,246,462,440,227,342,349,435,528, 547)

support    = c(156,100,160,254,194, 95,116,112,196,211, 328)
invalid    = c(  5,  9, 13, 27, 11, 20, 18, 28, 13,  9,  10)

valid = turnout - invalid
```

This data is reported by "the official electoral commission," allowing us to test the legitimacy of the election according to the government. From this, we calculate the invalidation rate and the support rate.

```
pInv = invalid/turnout
pSup = support/invalid
```

If there is no differential invalidation, any relationship between the invalidation rate and the position support rate should be due solely to expected random

---

[2] An invalidation plot is a scatter plot of the invalidation rate (vertical-axis) against the candidate support rate (horizontal axis).

variation in the election. This is tested using regression.[3] The results are as expected:

```
             Estimate  Std. Error  z value  Pr(>|z|)
(Intercept)  -2.5567   0.3824      -6.687   2.28e-11 ***
pSup         -1.2328   0.7698      -1.602   0.109
```

The relationship between the invalidation rate and the support rate is not stronger than one would expect in random chance

$$\text{p-value} = 0.1090 > 0.05 = \alpha$$

The p-value being greater than α means we should not reject the null hypothesis. This means that there is no statistical evidence of differential invalidation in this election. Because we created this election to be without differential invalidation, it is good to see that the test concludes the reality.

9.1.3 Differential Invalidation

Differential invalidation takes place when the ballots for one candidate or position are invalidated at a higher rate than for other candidates or positions. To simulate this, we can change some ballots from the "against" position to the "invalid" position, as in the following code.

```
diffInv  = c(10, 8, 7, 18, 24, 11, 21, 21, 23, 31, 21)

invalid2 = invalid + diffInv
valid2   = turnout - invalid2

pInv2 = invalid2/turnout
pSup2 = support/valid2
```

Note that this code increases the number of invalid ballots without changing the number of supporting ballots; it moves some votes from the "against" pile to the "invalid" pile (see Figure 9.4). In the presence of this differential invalidation, the regression conclusions change:

```
             Estimate  Std. Error  z value  Pr(>|z|)
(Intercept)  -1.3998   0.2788      -5.021   5.14e-07 ***
pSup2        -1.7936   0.5383      -3.332   0.000863 ***
```

The relationship between the invalidation rate and the support rate is *much* stronger than one would expect to arise from random chance ($p = 0.0009$). Thus, there is strong statistical evidence of a relationship between the invalidation rate and the position support. In other words, there is evidence of

---

[3] For consistency and speed, the regression method used in these examples is Binomial regression. This method will be covered fully in Section 9.2.3.

Differential Invalidation

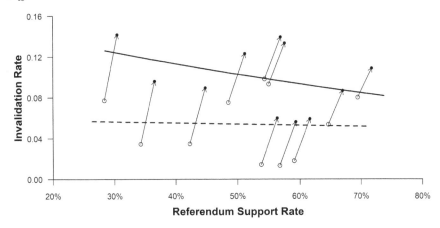

FIGURE 9.4
A second example of the effects of differential invalidation on the invalidation plot. Note that there is no significant slope in the free-and-fair election (dashed line), but there is when ballots are systematically invalidated based on who they were cast for (solid line).

unfairness in the election.[4] In fact, since higher support corresponds to fewer invalidated ballots, the differential invalidation *helped* the referendum support.

Beware

Note that the conclusion is *not* that there was cheating in the election. It is simply that there is evidence of unfairness in the election and that this unfairness benefited this particular position (or candidate).

While invalidation based on the ballot recipient is unfair, so too is a complicated ballot that causes those with poor eyesight, poor native-language skills, or poor patience to have difficulties properly completing it. If any of these items are significantly related to the position (or to the candidate), then the unfairness will also be shown in an appropriate invalidation plot.

---

[4] The effect of stuffing the ballot box with ballot papers in favor of the position is similar to that seen above. The support rate increases, while the invalidation rate decreases. Thus, the dots will be moved down and to the right.

9.2 Regression Modeling

Chapter 8 dealt with the case in which the government only reports the vote counts for the candidates at the electoral division level. There, we had to use digit tests to examine the likelihood that the data were generated from a "natural voting process."[Meb10] In addition to the candidate vote counts, governments may also report the number of invalidated ballots at the electoral division level. In such cases, we can perform an additional test, regression. This section briefly examines four types of regression.

The Free and Fair Hypothesis

In a democratic election, each person's vote counts the same. There are other requirements, but this is a necessary condition. In the presence of invalidation, the free and fair hypothesis reduces to each person's vote having the same *probability* of being invalidated as any other person's ballot.

From a statistical standpoint, this means that the invalidation must be independent of the candidate chosen on the ballot (or of the person voting).

The data model is
$$y_i = \beta_0 + \beta_1 x_i + \varepsilon_i$$

In other words, for division i, the observed invalidation rate (dependent variable, y_i) is the sum of the base invalidation rate, β_0, the effect of the position support rate (slope times independent variable, $\beta_1 x_i$) and unexplained variation, ε_i (the "residuals" or the "errors").

9.2.1 Ordinary Least Squares

A typical test for independence of two ratio-level variables is a t-test on the estimated slope parameter arising from ordinary least squares regression (OLS). The two most important strengths of using OLS to estimate the effect are that OLS is rather robust to violations of its assumptions and that OLS is easily performed and interpreted.

The objective of regression is to estimate β_0 and β_1 subject to some condition. For ordinary least squares regression, those estimators minimize the sum

Regression Modeling

of the squared errors. Math (see pages 146 and 224) provides the estimators:

$$\begin{cases} b_0 = \bar{y} - b_1 \bar{x} \\ b_1 = \dfrac{\sum (x_i - \bar{x}) y_i}{\sum (x_i - \bar{x})^2} \end{cases}$$

If we also make the usual assumption that the residuals are generated from a Normal distribution of mean zero and constant variance, $\varepsilon_i \sim \text{NORM}(0;\ \sigma^2)$, then we obtain the distribution of b_1 as

$$b_1 \sim \text{NORM}\left(\beta_1,\ \frac{\sigma^2}{\sum_{i=1}^{n}(x_i - \bar{x})^2}\right) \tag{9.1}$$

While the appendix to this chapter provides the proofs (page 225), the key here is that we are able to use probability statement 9.1 to obtain a p-value for the hypothesis that there was no differential invalidation in the election; that is for

$$H_0 : \beta_1 = 0$$

Example

To illustrate using ordinary least squares (OLS) regression, let us use the data from the previous examples.

```
modOLS1 = lm(pInv ~ pSup)
summary(modOLS1)
```

This code produces the following output.

```
              Estimate Std. Error t value Pr(>|t|)
(Intercept)    0.07279    0.03631   2.005    0.076 .
pSup          -0.05577    0.07022  -0.794    0.448
```

From this output, we see that the p-value is 0.4480. This value is sufficiently large (greater than $\alpha = 0.05$). As such, we should conclude that we did not detect differential invalidation; i.e., there is no significant evidence of it. This is as expected.

Now, let us use OLS on differentially invalidated data.

```
modOLS2 = lm(pInv2 ~ pSup2)
summary(modOLS2)
```

This code produces the following output.

```
              Estimate  Std. Error  t value  Pr(>|t|)
(Intercept)   0.17211   0.03522     4.887    0.000863 ***
pSup2        -0.15042   0.06501    -2.314    0.045954 *
```

From this output, we see that the p-value is 0.0460. This is sufficiently small (less than $\alpha = 0.05$). Because of this, we should conclude that there is some evidence of differential invalidation. This is also as expected.

The estimated slope (effect) is $b_1 = -0.15042$. Since this is negative, we can conclude that we have evidence that the differential invalidation helped the "in favor of" position on the referendum.

Significance: Statistical *vs* Practical

In the previous example, we determined that there was some evidence of differential invalidation. This came from the p-value being less than our selected $\alpha = 0.05$. In other words, we have *statistically* significant results. All this means is that there is enough evidence to conclude that the slope is not zero.

If all we care about is evidence for or against electoral unfairness, statistical significance is all we need to consider. If, on the other hand, we care about how extreme this unfairness is, we need to pay attention to the estimated slope and whether its value is really that important. This is a question of *practical* significance.

Ultimately, the statistician can only provide the level of statistical significance (p-value). It is up to the scientist/practitioner to determine if the estimated slope is important.

9.2.2 Weighted Least Squares

The two most important strengths of using OLS to estimate the effect are that OLS is rather robust to violations of its assumptions and that OLS is easily performed and interpreted. However, significant violations of the homoskedasticity (constant variance) requirement produce biased standard error estimates. This, in turn, means that the p-values are biased. Solutions to this issue include transforming the variables and adjusting the estimated standard errors.[Whi80, You79]

In lieu of adjusting the data or adjusting the standard error estimates, one should use knowledge about the data-generating process to improve the model. As the dependent variable is a proportion, the expected variance of the distribution of the observations under the null hypothesis is $\pi(1-\pi)/N_i$, where N_i is the number of votes cast in electoral division i. Note that the

Regression Modeling

variances are equal under the null hypothesis if *and only if* the number of votes cast in each division are equal. As turnouts tend to be unequal, the model will demonstrate natural heteroskedasticity. A usual solution to this issue is to use weighted least squares to estimate the slopes.

Recall that the objective of regression is to estimate β_0 and β_1 subject to some condition. For weighted least squares regression, those estimators minimize the *weighted* sum of the squared errors. Mathematics provides the estimators (see page 226). The following example shows how to use R to estimate β_1 using weighted least squares estimation.

9.2.2.1 Example

To illustrate using weighted least squares (WLS) regression, let us use the data from the previous example.[5]

```
modWLS1 = lm(pInv ~ pSup, weights=turnout)
summary(modWLS1)
```

This code produces the following output.

```
            Estimate Std. Error t value Pr(>|t|)
(Intercept)  0.06420    0.03632   1.767    0.111
pSup        -0.04741    0.07134  -0.665    0.523
```

From this output, we see that the WLS p-value of 0.523 is sufficiently large. Thus, we should conclude that there is no significant evidence of differential invalidation. Again, we are not surprised by this.

Now, let us use WLS on the differentially invalidated data.

```
modWLS2 = lm(pInv2 ~ pSup2, weights=turnout)
summary(modWLS2)
```

This code produces the following output.

```
            Estimate Std. Error t value Pr(>|t|)
(Intercept)   0.1641     0.0354   4.637  0.00123 **
pSup2        -0.1433     0.0663  -2.161  0.05895 .
```

From *this* output, we see that the WLS p-value is 0.05895. Thus, if we stick to the $\alpha = 0.05$ rule, we should conclude that there is no statistical evidence of differential invalidation in this election.

---

[5] In R, the weight parameter is proportional to the inverse of the variances. Thus, since the expected variance of the distribution of the observations under the null hypothesis is $\pi(1-\pi)/N_i$, and π is constant under the null hypothesis, we should weight on N_i, the turnout.

> **Significant or Not?**
>
> Note that we used two legitimate procedures to estimate the differential invalidation in this election. Note also that we came to two different conclusions—if we slavishly follow the $\alpha = 0.05$ rule.
>
> This should serve as both a warning and a clarification: Think about the results of these statistical tests simply as levels of evidence either for or against the null hypothesis (a free and fair election). The p-value can be thought of as the level of evidence in the data for the null hypothesis [as long as the test is appropriate].

9.2.3 Binomial Regression (GLM)

The previous two methods ultimately made the assumption that the dependent variable (invalidation rate) followed a Normal distribution, given the values of the independent variable (position support rate). Unfortunately, this is really not supported by the data.

An improvement on weighted least squares is Binomial regression, in which the dependent variable (*number* of invalidated ballots) follows a Binomial distribution, given the values of the independent variable (position support rate). Recall Chapter 2, where we first encountered the Binomial distribution. To follow a Binomial distribution, the dependent variable must meet these five requirements.

1. The number of trials, n, is known.
2. Each trial can be either a success or a failure.
3. The probability of a success is constant for the n trials.
4. The outcome of one trial is independent of the others.
5. The random variable is the number of successes.

In this context, the "number of trials" is the number of votes cast in that electoral division, and a "success" is defined as an invalidated ballot.

Note that there are two assumptions that are conflated in this setting: numbers 3 and 4. The fact that people with similar political beliefs tend to live near each other suggests that these assumptions, as a whole, may not be realistic.[MS04, WSS+01]

For instance, the probability of voting for a Republican in Utah will be higher than the probability of voting for a Republican in California. Similarly, the probability of voting for Ouattara was higher in the northwest than in

Regression Modeling

the southeast. This suggests two things: Geography needs to be considered *and/or* we need to relax requirements 3 and 4. We will address the second issue in the quasi-likelihood example.

Example: Binomial regression

To illustrate using Binomial regression, let us again use the data from the previous example.

```
depVar = cbind(invalid, valid)
modGLM1 = glm(depVar ~ pSup, family=binomial(link=logit))
summary(modGLM1)
```

This code produces the following output.

```
            Estimate Std. Error z value Pr(>|z|)
(Intercept) -2.5567     0.3824  -6.687 2.28e-11 ***
pSup        -1.2328     0.7698  -1.602    0.109
```

From this output, we see that the p-value of 0.109 is sufficiently large. Thus, we should conclude that there is no significant evidence of differential invalidation. No surprise.

Before moving on to analyzing the differentially invalidated data, let us return to the code and explain the three new parts. First, the regression function is glm instead of lm. The change in function reflects the change in the model estimation method. Previously, we have been estimating the slope (effect) using ordinary least squares (OLS). Once we moved to assuming the dependent variable follows a Binomial distribution (conditional on other specified values), we have moved into the realm of the generalized linear model, which is fit using maximum likelihood estimation (not OLS).[MN89, NW72]

In exchange for the increased flexibility, generalized linear models require one to explicitly specify three things: the linear predictor, the conditional distribution, and a function linking the linear predictor to the expected value of the distribution (called the link function). The **linear predictor** is the same as before. We are just giving is a fancy name, "eta"

$$\eta_i = \beta_0 + \beta_1 x_i \tag{9.2}$$

The **conditional distribution** for this model is the Binomial distribution. Note that the Binomial distribution requires specifying both the number of successes *and* the number of failures in each electoral division. This requirement gives rise to the line

```
depVar = cbind(invalid, valid)
```

This cbind function takes the number of invalid votes (successes) and the number of valid votes (failures) and binds them together as two columns in a matrix; hence, "cbind."

The **link function** needs to map the domain 0 to 1 (as required by the possible values of π) to the range $-\infty$ to ∞ (as required by the possible values of η in Eqn. 9.2). There are many functions that work here. The default in generalized linear models (GLMs) is the logit function,[6]

$$\text{logit}\, p := \log\left(\frac{p}{1-p}\right)$$

Finally, since we are using the logit link function, the estimated effect, $\hat{\beta}_1$, should not be interpreted in the same manner as in the classical linear model. While the direction (sign) and the p-values can be interpreted as usual, the magnitude corresponds to a different dependent variable. Thus, to obtain a graphic of the estimate, one will need to estimate the dependent variable in the GLM, then back-transform it with the inverse of the logit function to obtain estimated invalidation rates. However, if all you care about is whether there is evidence of differential invalidation, the p-value is all that is needed, as the following example shows.

```
depVar2 = cbind(invalid2, valid2)

modGLM2 = glm(depVar2 ~ pSup2, family=binomial(link=logit))
summary(modGLM2)
```

This code produces the following output.

```
            Estimate Std. Error z value Pr(>|z|)
(Intercept)  -1.3998     0.2788  -5.021 5.14e-07 ***
pSup2        -1.7936     0.5383  -3.332 0.000863 ***
```

Note that the p-value is very small. Thus, we could conclude that there is strong evidence of differential invalidation in this election. The negative sign on the estimate ($\hat{\beta}_1 = -1.7936$) indicates that this differential invalidation helped the supporters of the referendum.

---

[6] For an explanation of why the logit function is frequently the default, I refer you to the McCoullough and Nelder book that explains the mathematics behind generalized linear models.[MN89] Because there are so many other link functions for the Binomial distribution, I usually suggest using a few and seeing if the p-values vastly differ. They should not.

Is It *Really* a Binomial? Quasi-Likelihood

The Binomial distribution has a specific form with relatively little flexibility in its shape.[7] Because political views are "clumpy" (people who live near each other tend to have similar political views), it is quite likely that the data are "**overdispersed**." In other words, the variability in the data is too great to have come from a Binomial distribution. To detect this overdispersion, one can simply perform a chi-squared test on the reported residual deviance. The following line provides the upper 95% confidence critical value for the residual deviance, assuming no overdispersion.

```
qchisq(0.95, df=9)
```

This results in a critical value of 16.91898 (at the usual $\alpha = 0.05$ level). Since the residual deviance from this last model (20.311) is greater than this critical value, there is evidence of overdispersion in the model.

The adjustment for overdispersion is rather straight-forward. Instead of using the Binomial family in the function call, use the *quasi*binomial family:[8]

```
modGLM2q = glm(depVar2 ~ pSup2, family=quasibinomial(link=
    logit))
summary(modGLM2q)
```

This code produces the following output.

```
            Estimate Std. Error t value Pr(>|t|)
(Intercept)  -1.3998     0.4230  -3.309   0.0091 **
pSup2        -1.7936     0.8169  -2.196   0.0557 .
```

The p-value is no longer too small. As such, we should conclude that we did not detect differential invalidation in this election.

Skip the Binomial

Note that the overdispersion completely invalidates the Binomial model of page 216. Thus, any conclusions based on it are invalid. Overdispersion

---

[7]Given the number of votes cast, there is only one parameter for the Binomial distribution: π. Thus, it is more constrained than distributions with more parameters like the Betabinomial, which we will see later in Section 9.2.4.

[8]I am sure—and would surely *hope*—that statisticians are howling at this statement. As one can imagine, the reality is a bit different. By default, GLM estimates are determined using a method called "maximum likelihood estimation" (MLE). Using MLE requires the "dispersion parameter" to be 1 for the Binomial distribution. When overdispersion is detected, like here, the Binomial distribution is *not* the correct distribution. To adjust for this overdispersion, one can fit use a different estimation method: **maximum quasilikelihood estimation**. This adjustment is not perfect, but it is definitely better than doing nothing.

> has an effect similar to heteroskedasticity (non-constant variance). The standard errors are improperly estimated, meaning that the p-values are wrong.
>
> As a rule, I just use the quasibinomial family. If there *is* overdispersion, then this model adjusts for it. On the other hand, if there is *no* overdispersion, then this model is equivalent to the Binomial model.

9.2.4 Betabinomial Regression (VGLM)

The last regression method we will examine (and perhaps the best moving forward) is Betabinomial regression.[Yee15, YW96] The Betabinomial is a discrete distribution (like the Binomial) that has two additional parameters, allowing for greater flexibility in model fitting. It matches all of the strengths of Binomial regression and mitigates the possible drawback of overdispersion. The biggest drawback is that the mathematics become a bit more difficult to program. However, we are just practitioners, not computer scientists. We use what they make.

The most important part of performing Betabinomial regression is that we need to load a new package: VGAM.[Yee10] This package was developed by the creator/discoverer of the VGLM regression method, Thomas Yee.[9][Yee15]

The following code performs Betabinomial regression on the fair data.

```
modBB = vglm(depVar ~ pSup, family=betabinomial)
summary(modBB)
```

This code produces the following highly abbreviated output.

```
            Estimate Std. Error z value Pr(>|z|)
   pSup      -1.1035     1.5138  -0.729 0.466038
```

Again, the p-value for the estimated slope is rather large (p-value = 0.466038). Because of this, we have no real evidence of differential invalidation in this election. This conclusion matches previous analyses on this data, as well as our knowledge that the data were generated to represent an election with no differential invalidation.

The results for Betabinomial regression on the differentially invalidated data are

---

[9]To use this package, it must be installed on your computer, then "activated." The following code will install it (you need to do this only once on each computer you use): install.packages(VGAM). This code will activate it, which you need to do each session you use it: library(VGAM).

Regression Modeling

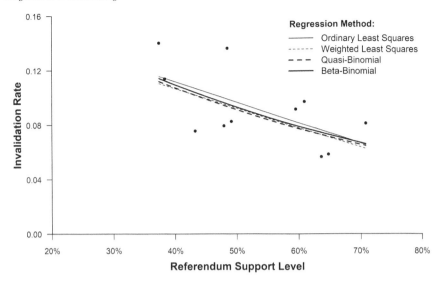

FIGURE 9.5
Graphic comparing the regression curves for the four methods discussed here. Note that they are all about the same.

```
              Estimate Std. Error z value Pr(>|z|)
    pSup2     -1.7901    0.7246   -2.470  0.013494 *
```

Again, note that there is some evidence of differential invalidation that helped the supporters of the referendum (p-value = 0.013494; $\hat{\beta}_1 = -1.7901 < 0$).

9.2.5 Comparing the Models

In the previous four sections, we looked at four different modeling schemes. Each one has its strengths and its weaknesses. The main strength of OLS is that we have all experienced it and understand it. Both OLS and WLS provide regression output that is easily interpreted in terms of the slope. Binomial regression improves upon OLS and WLS in terms of modeling the type of dependent variable more correctly. Betabinomial regression improves upon Binomial regression in that it does not need an adjustment for overdispersion; the Betabinomial distribution is much more flexible than the Binomial.

Figure 9.5 shows all four models on the same graphic. The most stunning take-away is that all four models provide roughly the same *predictions*. This should make us feel good, since all four methods tell the same story about the data. This is also supported by the p-values provided in the table below.

Method	Free-and-Fair	Differentially Invalidated
Ordinary Least Squares	0.448	0.0460
Weighted Least Squares	0.523	0.0590
Quasi-Binomial	0.519	0.0557
Betabinomial	0.466	0.0135

The middle column consists of the p-values for the free and fair election; the right column, for the differentially invalidated data. All of the models provide roughly the same p-values for each of the two scenarios. In all models, there is no evidence of differential invalidation in the free-and-fair election. For the differentially invalidated election, only the Betabinomial model indicated anything more than very weak evidence for differential invalidation.

The Final Upshot

For actual election analysis, I only use the Betabinomial regression. Its requirements fit the election models much more closely than the others in the list. However, I would still conclude that there is not significant evidence of differential invalidation. While the p-value is less than our usual $\alpha = 0.05$, I am uncomfortable attacking a government like this. In electoral forensics, I select $\alpha = 0.005$ for this very reason. That, and I do not want those governments coming after me.

9.3 Examining Côte d'Ivoire

Let us apply the techniques of this chapter to the official results of the 2010 presidential election in Côte d'Ivoire. These data were published by the Independent Electoral Commission (CEI).[Ind10] Thus, they are official government data.[10]

Figure 9.6 is the invalidation plot. Note that all four models fail to suggest any differential invalidation because all regression curves are relatively flat. The following table echoes those results.

---

[10] The script for this analysis can be found in this chapter's appendix, page 227.

Examining Côte d'Ivoire

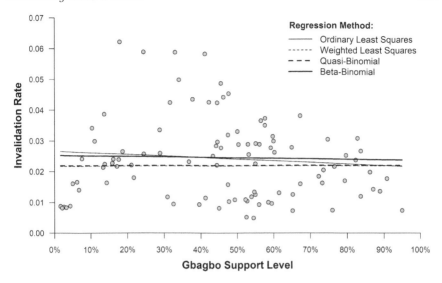

FIGURE 9.6
Graphic comparing the regression curves for the four methods discussed here for the second round 2010 presidential election in Côte d'Ivoire. Note that they are all about the same, meaning that they are all telling the same story about the election.

Method	p-value
Ordinary Least Squares	0.340
Weighted Least Squares	0.995
Quasi-Binomial	0.995
Betabinomial	0.745

Recall from the opening narrative that the 2010 presidential election was disputed between the two candidates. The CEI released results showing that Ouattara won. However, the Constitutional Court declared that those results were tainted with fraudulent votes. As a result the Constitutional Court threw out the votes from several divisions and declared Gbagbo the winner.[BBC11a, Yì0]

This analysis, however, does not show any significant evidence of differential invalidation in the CEI data. Note that this test is only of a specific type of unfairness and that it is not designed for other types. The code for this analysis is in the appendix, Section 9.6.5.

9.4 Conclusion

Last chapter, we examined a few test of unfairness when the government only provided the vote counts for the candidates. In this chapter, we looked at more powerful tests that can be used when the government provides the invalidation rate.

In all of the differential invalidation procedures, we tested for a relationship between the invalidation rate and the proportion of the vote in favor of a position or a candidate. This led to several available regression tests, each with their advantages and their disadvantages. Ultimately, it appears as though the quasiBinomial model and the Betabinomial model are the best options, with the Betabinomial VGLM model most likely being the better of the two.

Next chapter, we will be examining geographic tests. Note that these tests will require even more data from the government. That new data is the boundary of each electoral division.

While the next chapter is still tied to regression, it does introduce additional independent variables. These new variables represent the location of the electoral division, usually latitude and longitude.

Plus, there are some great maps that tell tales of their own!

9.5 Extensions

In this section, I provide some practice work on the concepts and skills of this chapter. As usual, some of these have answers, while others do not.

Review

1. Compare and contrast the four regression methods discussed in this chapter.
2. What is differential invalidation?
3. What is ballot box stuffing?
4. How does differential invalidation affect the electoral divisions displayed on an invalidation plot.

Chapter Appendix

5. Compare and contrast statistical significance and practical significance.

Conceptual Extensions

1. Why do the conclusions in this chapter never state that an election was unfair?
2. How does ballot box stuffing affect the electoral divisions displayed on an invalidation plot.
3. What does a p-value indicate?
4. If you is concerned with practical significance, should you pay attention to the confidence interval or the p-value?

Computational Extension

1. Perform the generalized Benford test on the Côte d'Ivoire data to see if there is evidence of vote count-fraud in the 2010 election.

9.6 Chapter Appendix

As usual, the appendix to the chapter holds information that I feel detracts from the flow of the writing. Here, I think that these proofs are absolutely important to those who are interested in how the statistical techniques are based on the assumptions we make and the mathematical results of those assumptions.

9.6.1 R Functions Used

A statistical program is invaluable in gaining a better understanding of statistics and its/their relationships to the world around us. The following R functions were featured in this chapter.

- `cbind(a,b,c,...)`

  This function binds the vectors into a single matrix by columns. There is a sister funct, `rbind`, that does this by rows.

- `glm(y~x, family)`

This function performs maximum likelihood estimation (MLE) of the variable y depending on x. The family specifies the distribution of y conditioned on x. Note that using family=gaussian will provide the same estimates as doing OLS. For this discipline, family=binomial is the proper choice if there is no overdispersion and family=quasibinomial if there is. If the link function is not specified, then it is assumed to be the logit link, which is as good as any.

- lm(y~x)

This function performs ordinary least squares (OLS) estimation of the variable y depending on x.

- lm(y~x, weights)

This function performs weighted least squares (WLS) estimation of the variable y depending on x, where the data are weighted according to the weights. These weights are proportional to the precision (the inverse of the variance).

- qchisq(p,df)

This provides the pth quantile of a Chi-square distribution with df degrees of freedom. This is used to test for overdispersion. However, if you use the Betabinomial family, then there is no need to test for overdispersion. The distribution does not allow for it.

- summary(m)

This provides the regression table (and other items) in a nice output. It should alwyas be run on the model to give the results you seek.

- vglm(y~x, family)

This function performs maximum likelihood estimation (MLE) of the variable y depending on x. The family specifies the distribution of y conditioned on x. For this discipline, family=betabinomial is the proper choice. This distribution contains the Binomial as a simple case. Thus, there is no reason to use anything other than this function.

9.6.2 Proof of OLS Estimators, II

Back in the appendix to Chapter 6 (page 146), we saw one proof of the OLS estimators. It relied on simple linear regression, one dependent variable and one independent variable.

Matrices, being a summary of the arithmetical structure, frequently makes the problems easier to solve in general. In matrix form, the linear model can be written as $\mathbf{Y} = \mathbf{XB} + \mathbf{E}$, where $\mathbf{Y}$ is the column vector of the dependent variable values (invalidation rates), $\mathbf{X}$ is the design matrix consisting of a column of 1's and a column of values for each independet variable (here, just of candidate support rates), $\mathbf{B}$ is the vector of effect sizes, and $\mathbf{E}$ is the vector of residuals (the deviations not explained by the model).

The OLS estimators are again determined using the same calculus ideas applied to minimizing the objective function $Q = \sum_{i=1}^{n} e_i^2$. In matrix form, this objective function is just $Q = \mathbf{E}'\mathbf{E}$:

$$\begin{aligned} Q &= \mathbf{E}'\mathbf{E} \\ &= (\mathbf{Y} - \mathbf{XB})'(\mathbf{Y} - \mathbf{XB}) \\ &= \mathbf{Y}'\mathbf{Y} - \mathbf{B}'\mathbf{X}'\mathbf{Y} - \mathbf{Y}'\mathbf{XB} + \mathbf{B}'\mathbf{X}'\mathbf{XB} \end{aligned}$$

Note that each term is a scalar. Thus, each term equals its transpose. This observation leads to this simplification:

$$Q = \mathbf{Y}'\mathbf{Y} - 2\mathbf{B}'\mathbf{X}'\mathbf{Y} + \mathbf{B}'\mathbf{X}'\mathbf{XB}$$

Now, we take the derivative of this function with respect to the vector $\mathbf{B}$:

$$\frac{d}{d\mathbf{B}} Q = -2\mathbf{X}'\mathbf{Y} + 2\mathbf{X}'\mathbf{XB}$$

Setting this equation to zero and solving for b, the OLS estimator gives

$$\begin{aligned} 0 &= -2\mathbf{X}'\mathbf{Y} + 2\mathbf{X}'\mathbf{Xb} \\ \mathbf{X}'\mathbf{Xb} &= \mathbf{X}'\mathbf{Y} \\ \mathbf{b} &= (\mathbf{X}'\mathbf{X})^{-1}\mathbf{X}'\mathbf{Y} \end{aligned}$$

Note that this matrix solution works for *any* number of independent variables, not just one; it is much more general and helpful. Also, it is what the computers do "under the hood" when calculating OLS estimators.

9.6.3 OLS Estimator Distribution

To move beyond the mathematics and into the statistics, we make the assumption that $\varepsilon_i \sim \text{NORM}(0;\ \sigma^2)$. Since $y_i = \beta_0 + \beta_1 x_i + \varepsilon_i$, and since the only random variable on the righthand side is ε_i, we conclude $y_i \sim \text{NORM}(\beta_0 + \beta_1 x_i;\ \sigma^2)$.

Recall the formula for the OLS estimator of b_1 is

$$b_1 = \frac{\sum_{i=1}^{n}(x_i - \bar{x})y_i}{\sum_{i=1}^{n}(x_i - \bar{x})^2}$$

As the x_i are not random variables, and as a linear combination of Normal random variables is a Normal random variable, it is easily shown that

$$b_1 \sim \text{NORM}(\beta_1;\ \sigma^2/S_{xx})$$

Here, as elsewhere, $S_{xx} := \sum_{i=1}^{n}(x_i - \bar{x})^2$.

The next step is to create a test statistic that is a function of the data, of β_1, of nothing else unknown, and has a distribution that does not depend on the parameter β_1.[RT14, Sha07, WMS08, see *pivotal quantity*] A first try would be the obvious z-transform [Aho14]:

$$\frac{b_1 - \beta_1}{\sqrt{\sigma^2/S_{xx}}} \sim \text{NORM}(0; 1)$$

Unfortunately, the value of σ^2 in the denominator of the statistic (the realistic variability in the population) is unknown. As such, this test statistic will not work. However, under the Normality assumption made above, we know another important distribution:

$$\frac{(n-2)}{\sigma^2}\text{MSE} \sim \chi^2_{n-2}$$

That is, the mean squared error (MSE $:= \frac{1}{n-2}\sum e_i^2$), properly scaled, follows a Chi-squared distribution with $n-2$ degrees of freedom.

Why is this useful? We can use the definition of the Student's t distribution [Gos08, RT14, WMS08] and the above distributions to come to an acceptable test statistic:

$$\frac{b_1 - \beta_1}{\sqrt{\text{MSE}/S_{xx}}} \sim t_{n-2}$$

With this, we have our formulas for the endpoints of a confidence interval and for p-values.

9.6.4 WLS Estimator and Distribution

In terms of the current paradigm, the model remains $\mathbf{Y} = \mathbf{XB} + \mathbf{E}$. However, we are allowing for heteroskedasticity: $\mathbf{E} \sim \text{NORM}(\mathbf{0}, \sigma^2 \mathbf{D})$, where $\mathbf{D}$ is an arbitrary, yet known, diagonal matrix.[Chr02] To obtain the least squares estimators under this new error distribution, we note that we can pre-multiply by the square root of $\mathbf{D}$ and return to the world of OLS, because

$$\mathbf{E} \sim \text{NORM}(\mathbf{0}, \sigma^2 \mathbf{D}) \iff \mathbf{D}^{-1/2}\mathbf{E} \sim \text{NORM}(\mathbf{0}, \sigma^2 \mathbf{I})$$

This means that the matrix equation

$$\mathbf{D}^{-1/2}\mathbf{Y} = \mathbf{D}^{-1/2}\mathbf{XB} + \mathbf{D}^{-1/2}\mathbf{E}$$

which can be rewritten (to make the point) as

$$\mathbf{Y}^* = \mathbf{X}^*\mathbf{B} + \mathbf{E}^*, \text{ with}$$

Chapter Appendix

$$E^* \sim \text{NORM}(\mathbf{0},\ \sigma^2 \mathbf{I})$$

is just the matrix equation for OLS on the transformed data. This quickly leads to the following WLS estimator for **b** (from Section 9.6.2, above).

$$\begin{aligned}
\mathbf{b} &= (\mathbf{X}^{*\prime}\mathbf{X}^*)^{-1}\mathbf{X}^{*\prime}\mathbf{Y}^* \\
&= \left(\mathbf{X}'\mathbf{D}^{-1/2}\mathbf{D}^{-1/2}\mathbf{X}\right)^{-1}\mathbf{X}'\mathbf{D}^{-1/2}\mathbf{D}^{-1/2}\mathbf{Y} \\
&= (\mathbf{X}'\mathbf{D}\mathbf{X})^{-1}\mathbf{X}'\mathbf{D}\mathbf{Y}
\end{aligned}$$

This leads to at least one important question: What is **D** in our current context? We decided in Section 9.2.2 that the number of invalidated ballots follows a Binomial distribution. This means that the inherent variance (variability, randomness) of the invalidation rate in electoral division i is $\sigma_i^2 = \pi(1-\pi)/N_i$. And so, the variance of the **E** vector is

$$\mathbb{V}[\,\mathbf{E}\,] = \sigma^2 \mathbf{D}$$

$$= \begin{bmatrix}
\pi(1-\pi)/N_1 & 0 & 0 & \cdots & 0 \\
0 & \pi(1-\pi)/N_2 & 0 & \cdots & 0 \\
\vdots & \vdots & \vdots & \ddots & \vdots \\
0 & 0 & 0 & \cdots & \pi(1-\pi)/N_n
\end{bmatrix}$$

$$= \pi(1-\pi) \begin{bmatrix}
1/N_1 & 0 & 0 & \cdots & 0 \\
0 & 1/N_2 & 0 & \cdots & 0 \\
\vdots & \vdots & \vdots & \ddots & \vdots \\
0 & 0 & 0 & \cdots & 1/N_n
\end{bmatrix}$$

This last matrix is the **D** matrix in this context.

9.6.5 Côte d'Ivoire Analysis Code

```
### Preamble
dt = read.csv("civ2010pres2cei2.csv")
attach(dt)

library(VGAM)


## Calculations
pInv = Nuls/Votants
pGba = GBAGBO/Suffrages
depVar = cbind(Nuls,Suffrages)
```

```
### Modeling

## OLS
modOLS = lm(pInv~pGba)
summary(modOLS)


## WLS
modWLS = lm(pInv~pGba, weights=Votants)
summary(modWLS)


## Quasi-Binomial
modBIN = glm(depVar~pGba, family=quasibinomial(link=logit))
summary(modBIN)


## Betabinomial
modBB = vglm(depVar~pGba, family=betabinomial)
summary(modBB)
```

10

Polling Analysis: Brexit Vote, 2016

The past chapters have presented a lot of polling information. This chapter applies it all to a single referendum: The June 2016 Brexit vote. Brexit refers to the United Kingdom's decision to leave the European Union (EU) following a referendum. The UK was a member of the EU for over 40 years, and its departure marked a significant shift in the country's political and economic landscape. Brexit has been driven by various factors, including concerns about sovereignty, immigration, and regulation, as well as dissatisfaction with the EU's bureaucracy and perceived democratic deficit.

This chapter raises a lot of questions—some of which are undecided in polling research. However, it should give you a taste of what poll analysts do to produce their estimates.

United Kingdom: The Brexit Vote of 2016
Strictly speaking, the European Union (EU) came into being with the ratification of the Maastricht Treaty of 1993. However, the beginning of the EU goes at least as far back as the 1951 Treaty of Paris, which created the European Coal and Steel Community (ECSC). This union regulated the industrial production of the six signatories with an intent to reduce the probability of a war within Europe, specifically between France and Germany. In addition to the ECSC, the Treaty of Rome (1957) created the European Economic Community (EEC), which united the economies of the signatories even further. The goal of this was to integrate the economies of Europe, thus creating a common market and greater wealth in Europe.

In 1961, the United Kingdom applied to join the EEC. However, French President Charles de Gaulle blocked British entrance in an effort to reduce American influence. When Pompidou succeeded de Gaulle as French President, the British began their application anew. Finally, on January 1, 1973, the United Kingdom joined the EEC.

However, when the European Union was formed in 1993, it consisted of more than just the European Economic Community. The three pillars of the EU also included the Common Foreign and Security Policy (taking care of foreign policy and military matters) and the Police and Judicial Co-operation in Criminal Matters (taking care of domestic criminal matters). Thus, while Britain did hold a referendum on remaining in the EEC (in 1975), it never held a referendum for joining the European Union.

Since 1993, various factions of various British parties have pushed for the United Kingdom to withdraw from the EU. Arguably, the most vocal Euroskeptic party was the UK Independence Party, which formed in 1991, changed its name in 1993, and increased its vote share in Parliament in each election until after the Brexit vote in 2016. Its vote share of 12.6% in 2015 even ensured that it was a member of the opposition. However, the success of UKIP was more a matter of the rising Euroskeptic tide in Britain. According to Curtice, support for Britain leaving the EU trended upwards from 1993 onwards, from 38% in 1993 to 65% in 2015, culminating in the Brexit Referendum of 2016.[Tar16]

On June 9, 2015, the House of Commons voted 544-53 in favor of holding a *non-binding* referendum on whether the United Kingdom should remain in the European Union. The vote, which was held on June 23, 2016, asked:

> Should the United Kingdom remain a member of the European Union or leave the European Union?

Throughout the year-long season, the opinion polls rarely showed the "Leave" position ahead. However, on June 23, the voters supported the Brexit position by a vote of 51.89% to 48.11%. As Figure 10.1 suggests, the support for Brexit tended to be in England and Wales. Scotland and Northern Ireland tended to support Britain's continued existence within the European Union (as did the overseas territory of Gibraltar).

Beyond the economic and social shock waves of this surprising result, this was "yet another" miss for polls. Many more began to wonder if opinion polls were as valuable as they once seemed. And so, for this chapter, we will closely examine the 2016 Brexit Referendum in the United Kingdom, using it to illustrate the techniques of the past few chapters, as well as to teach us some lessons about polling—especially about stratified sampling.

10.1 Know Your Data

The first step in any analysis is to get to know your data. Look and notice things. Make graphics and study them for abnormalities and patterns. Make a map and think about the geography. Know your data.

Know Your Data 231

FIGURE 10.1
Map of the Brexit support in each of the United Kingdom's four constituent countries. The overseas territory of Gibraltar is included in England. The data are from Wikipedia.[Wi20c]

Figure 10.2 shows the results of all polls in the data set, along with a smooth loess curve designed to show trends. Note that from early April onward, support for Brexit increased a bit in the polls, but plunged in the last couple days. Was this plunge real or an artifact of the samples taken?

Recall that polls differ in many important aspects. For instance, some polls are online, while others use the telephone. Does this difference matter? Another possibly important difference is in who is kept in the sample. Should it be all adults, or should the polling firm limit the sample to those who are going to vote in the election? If the latter, how can the polling form determine who will vote?

In general, we would expect both the mode and the sampled population to matter. But, that is just an expectation. Does it seem to matter in this election?

10.1.1 Telephone vs. Online

The 127 polls in our sample consisted of a mixture of online (85) and telephone (42) polls. A persistent question concerns whether or not these two

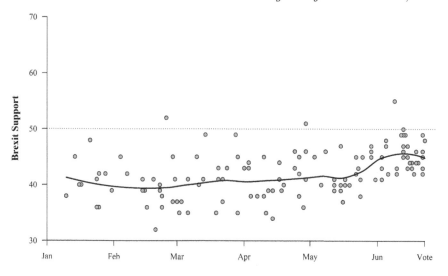

FIGURE 10.2
A plot of all polls in the data set, with a smoothing curve shown. The actual result for the 2016 Brexit referendum is 51.89% in favor. The bulk opinion polling data are from Wikipedia.[Wi19]

modes are unbiased. Figure 10.3 divides the polls into their two modes and plots smoothing curves for each subset. Note that the telephone polls were much more volatile, especially from early April onward. The online polls were much more stable. While one would expect this to be true for individual polls, since the sample sizes of online polls tends to be much larger than those of telephone polls, it is the polling averages that varied. This hints at a fundamental difference in the sampled populations attracted to each of the two polling modes.

Why might the two sampled populations be different? Was it a matter of the young vs. the old? the females vs. the males? Or, is it something else, something more fundamental? Current thought is that there are two primary sources of error. The first is the quality of online polling. The second is the different populations being sampled. The reality is that online polling, while better than it was a decade ago, is still not as statistically sound as telephone polling.[Coh19] However, a separate analysis finds that the bulk of the discrepancy between the two mode is due simply to Internet and phone polls sampling different people.[Sin16]

Because of this difference, should the online polls be eliminated from consideration? Should the telephone polls carry a higher weight than the online polls? Pew Research studied online polling methods in 2016 and found that their accuracy rates ranged from 66% to 76%.[KMK+16] This large range

Know Your Data   233

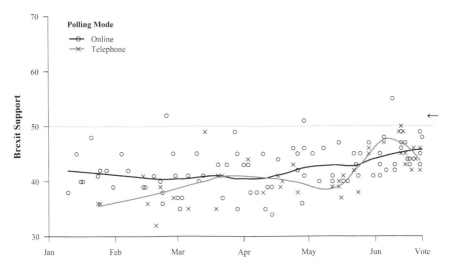

FIGURE 10.3
A plot of all polls in the data set, separated by mode of contact: online or telephone. The arrow, ←, is the actual result for the 2016 Brexit referendum.

indicates that some online firms do better than others. Perhaps we should weight according to the polling firm instead of the poll.

10.1.2 The Population of Interest

For estimating the outcome, it is very clear that the population of interest is all people who actually voted on June 23 in the Brexit referendum. It is also clear that this target population does not exist until *after* the polls close (Section 1.2.1). How can we possibly estimate the average position of a non-existent population? This is one of the most difficult challenges in polling.

Some polling firms tend to estimate this future population using some method, while others simply report estimates on the current level of support. To estimate who will be a member of the voting population, some firms simply ask the respondent if they plan to vote. Others provide a 0–10 scale and weight the results according to that value. Still others use proprietary algorithms that are based on such factors as previous voting patterns and the importance of the vote results.

Unfortunately, many polling firms do not indicate who is in the sampled population, whether there is weighting involved, how those weights are determined, or how they are used. In several cases, the firm provides this information, but the media reports do not. This makes investigating the effect of the sampled population difficult.

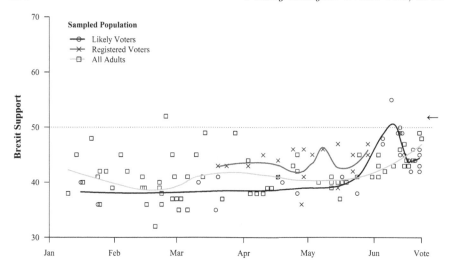

FIGURE 10.4
A plot of all polls in the data set, separated by population: likely voters, registered voters, all adults. The arrow, ←, is the actual result for the 2016 Brexit referendum.

To gather the data used in Figure 10.4, I tried to determine the sampled population used from the polling firm's report. In some cases, the report was very explicit; in others, less so. When a polling firm tended to one type of population, I would impute that value. If I had no idea, I would leave the poll blank.

The interesting thing about this graphic is that polls of "likely voters" became very erratic in the waning weeks of the election period. The position of "all adults" was much more stable. Furthermore, the "all adult" surveys came closer to the actual Brexit vote (the left arrow, ←).

What could cause this? Did the "likely voter" polls severely underestimate who would vote? That seems to be the logical conclusion from Figure 10.4. This may be due to a couple of reasons: those who said they would vote decided to stay home; or those who said they would not vote decided to vote.

Or did the undecided voters "*break*" in a particular direction?

10.1.3 Breaking Undecideds

Figures 10.2 through 10.4 show the proportion of those polled who support Brexit. This actually misstates the support levels in the population because the "undecideds" are not counted. Figure 10.5 shows a better picture of what the polls are really saying.

Know Your Data   235

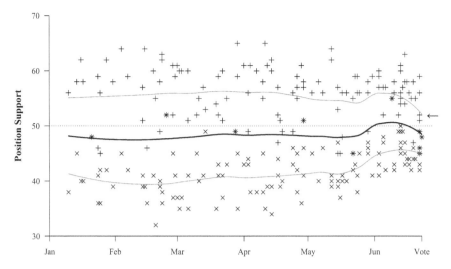

FIGURE 10.5
A plot of all polls in the data set, with a smoother curve. The lower curve is the level of support for Brexit in the poll. The upper curve is the level of "maybe-support" for Brexit. The arrow, ←, is the actual result for the 2016 Brexit referendum.

Like the other figures, Figure 10.5 provides the support for Brexit in each poll (×). It also provides the level of "maybe support" for Brexit (+).[1] The two grey loess curves essentially illustrate the upper and lower bounds on Brexit support. The dark smoother curve is the middle of the two extremes, representing the most likely level of support.

Examining the polls in these terms does a much better job of illustrating the level of uncertainty caused by the "undecided" respondents. The upper bound for voting day is 52.18%; the lower, 44.87%. The observed value is 51.89%, which is between these two extremes.

While the final interval does contain the true vote share, it is still a rather wide 7.31%. It would be nice to increase the precision of our estimate (reduce the interval width) while retaining the expected accuracy. The next section seeks to accomplish these two goals by combining several polls.

---

[1]This "maybe support" is the proportion of respondents in favor of Brexit *plus* the proportion of undecided respondents in the poll. In other words, if the 50% line is in the space between the two curves, the polls are uncertain as to which side will win. However, if the line is either entire above the upper curve or entirely below the bottom curve, then the polls are either concluding that the vote will be in favor of Brexit (above) *or* against Brexit (below).

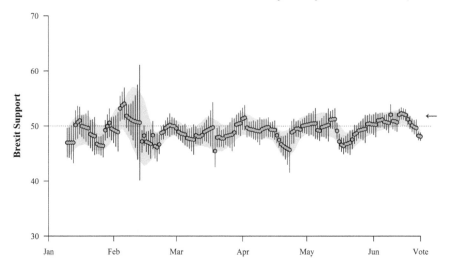

FIGURE 10.6
A plot of the estimated support for Brexit. The vertical segments indicate the 95% confidence interval for each day. The grey shape is the smoothed confidence interval envelope. The arrow, ←, is the actual result for the 2016 Brexit referendum.

10.2 Combining the Polls

Recall that Chapter 6 took us beyond estimating support from just one poll. One could combine polls to obtain better estimates of the election outcome. All that was needed was to select an appropriate weighting function and perform the calculations.

Figure 10.6 shows one set of daily estimates. For this graphic, I removed the "undecided" from consideration. I then used the Gaussian weighting function with a half-life of $\theta = 3$ to make use of previous polls. This choice came from personal experience and nothing particular to this election. Such a small half-life allows the model to better reflect a sudden change in referendum support. The result of allowing this is that the amount of data is reduced. Ultimately, the choice of θ is a balancing act between model responsiveness (smaller theta) and efficiency (larger theta).

In addition to weighting polls according to their age, this model also weighted polls according to the researcher's belief in their quality. While this is fundamentally a subjective activity, and two researchers will most likely weight the same poll differently, it does allow the researcher finer control over what is included in the model, and how much it counts. In accord with

Section 6.4, the effect of the weighting is only on the sample size. Thus, if I give the XYZ poll a weight of 0.50, then I would simply multiply its sample size by 0.50 to obtain the weighted poll result. For this particular exercise, I weighted the polls with values between 0.50 and 1.00 based the information present and whether it was a telephone poll.

Note that this model concludes that the majority of voters will vote *against* Brexit, which is not what happened. It does, however, reflect all of the other models used, as well as expert opinion at the time.[Erl16] Had I weighted the online polls higher, and the telephone polls lower, the final estimate would have been closer.

However, the problems with the Brexit polling were much deeper and more fundamental, as the next section discusses.

10.3 Discussion: What Went Wrong?

The results of the Brexit vote were surprising to many. The markets reacted negatively to the news that the United Kingdom was starting to undo two generations of work to bring Europe closer together. The stark difference in Brexit support between England and Scotland led Holyrood to declare that Scottish independence would be voted upon in the near future (see Chapter 4).[Erl16] In short, we were wrong.

But *why* were we wrong?

10.3.1 Stratified Sampling and Demographics

Recall Chapter 4, where we discussed stratified sampling. This is the technique used by polling firms to adjust the demographic bias of a sample. In doing this, the estimates tend to be closer to reality.

Section 4.1 warns us, however, that the stratified estimator is biased unless the assumed population proportions (weights) are correct. In Section 4.2.2, we performed some statistical experiments to see how close we had to be with our weights before the estimates became so poor that they were "worthless." For the experiment described, we concluded that being within 2 percentage points led to advantages over simple random sampling.

However, that does not answer our *real* question: How far off can we be before our support estimates are worthless? In the Brexit example, one could easily conclude that predicting the wrong result makes the prediction worthless. Since the estimates were so close to 50%, the weights had to be even closer to reality than the 2pp rule of thumb (page 84).

Add to this the fact that the population weights refer to the demographics of those who actually turned out to vote. If the turnout is different than what is expected, then the weights are necessarily wrong. Is this what we find in the Brexit referendum? Yes.

The 72% turnout was the highest for any British election since 1992, and for any British referendum *ever*.[Erl16] This high turnout was not constant across all groups. The older voters turned out at a much higher rate than did the younger, 90% of the former vs. 64% of the latter.[Hel16] This, in itself, is not a problem. It becomes a problem when age is correlated with referendum support, and the older voters tended to support Brexit while the younger did not.[Moo16] Since the polling firms underestimated the turnout for older people, they underestimated the total support for Brexit (see Section 4.2.1).

Who We Missed

Furthermore, it appears as though certain demographics, usually not weighted for in stratified sampling, were important in predicting the voter's intent. A person's level of education was highly correlated with the person's support for Brexit. Of those with a university degree, 32% voted for Brexit, while those without college experience[2] voted 70% for Brexit.[Moo16]

That age is related to Brexit support is not problematic—unless the weights used in adjusting the individual samples are not correct. That education level is related is also not a problem, unless the distribution of education is related to turnout, in which case it is a huge problem. It is even more of a problem if those last-minute deciders tended to be in the same educational category.

It is quite reasonable to expect that those with a university degree to tend to make decisions earlier, since they have much more access to news. However, to determine if the undecided voters tended to be correlated with education level, one would need the data.

Unfortunately, education level tended to not be stratified by polling houses. Age was. Gender was. Political orientation was. Education level was not.

I believe *this* was the main reason the polls were wrong.

Detecting this problem is straight-forward. Polls just have to ask a question about educational attainment. Then, when they start to detect a relationship between education and Brexit support, they need to start stratifying on education. However, what other demographics need to be recorded? Current

---

[2]Technically, the group supporting Brexit at 70% is those with a GSCE or less. Those with an A-level (or equivalent) were evenly split on Brexit, as were those with some university.[Moo16]

polls can take upwards of 10 minutes—or more—to complete. Adding additional questions will only drive that time up, and the response rates down. As with most things in statistics, increasing quality in one area tends to reduce quality in another.

10.4 Conclusion

The main purpose of this chapter was to illustrate the theory of the past few chapters with a concrete application. The 2016 Brexit vote was selected because it had a lot of data *and* because the polls "missed" the right answer. This allowed us to investigate more fully what went wrong and make suggestions for future polling.

Brexit opinion polls tended to use stratified sampling. They did this to help adjust for the demographic randomness in each sample. That is, while a single random sample may easily have a gender-ratio of 1.35 (135 females per 100 males responding), skewing the estimate toward the "female position," stratified sampling would weight the responses so that the sample would effectively have a ratio of 1.00 (or whatever the polling house believed the voting population would be). In this election, there was not such a gender gap. Females and males tended to support Brexit at about the same rate. The last opinion poll by YouGov had females support Brexit at 48%; males at 49%.[You16]

Thus, it was not the gender gap that caused problems for the Brexit polls. The problem turned out to be age and education level. While the polls *did* tend to stratify on age, they did so at the wrong level. Older voters turned out at 90%, much higher than weighted for in the polls. Furthermore, there was a strong differential support for Brexit across the education categories, which tended to not be a stratifying variable.

The Lessons of Brexit

The Brexit vote took place eight years ago. My first version of this chapter was *four* years ago. Time has passed, but it appears as though the lessons remain unlearned for a lot of us.

If there is a lesson to Brexit for *pollsters*, it is that we need to be more aware of how certain demographics will support (or not support) the referendum position. Once we are aware of a confounding variable, we need to stratify on that variable. This will be difficult at first, especially since we will not know the voting turnout for each of the strata in the upcoming election. However, it will be a necessary first step.

If there is a lesson to Brexit for the *media*, it is that more information needs to be presented to our audiences about the polls. Merely saying that Poll X estimates 48% support for Brexit ignores a lot of important information. The sample size, the strata, and the sampled population are key pieces of data that will help your readers better understand the article. Furthermore, ensuring that the polling house is able to defend their choices and give some indication of how important their assumed weights are to their final estimates would allow your readers to know where to focus.

Finally, if there is a lesson to Brexit for the *voters*, it is this: **vote**.

11

Election Analysis: Collier County

One thing I love about being a college professor is that I get to work with students, other faulty members, and—most importantly—members of the community. Those unexpected telephone calls on Tuesday afternoons have never failed to result in something interesting for me. This chapter arose from such a phone call.

In an initial analysis of the election results in Collier County, Florida, odd patterns were seen. For the next couple of months, I examined the election and explained the patterns. This chapter is an edited version of the resulting report. Use it to see how we can use statistics to better understand elections.

Introduction

One of the members of the Republican Executive Committee in Collier County, Florida, contacted me to have me analyze the results of the 2020 presidential election in the county. The three members of the executive committee had already received the necessary data from the Collier County Supervisor of Elections and started to analyze it. Odd patterns in the data appeared that rightly demanded to be explained. At this point, they decided to contact an expert on the matter to get better insight into what happened with the election results.

With that, Terry found my number and gave me a call. After a discussion, she sent the Cast Vote Record (CVR) files to me for analysis. The ultimate concern was that the optical scanner machines switched votes from one candidate to the other. Unfortunately, this question cannot be answered with the available data. To answer this question, one must know the original vote and compare it to the reported vote. Since the United States uses the Australian ballot in elections (a.k.a. the "secret ballot"), the original vote is unknowable. As such, it is impossible to test the machines with this data.

In lieu of answering that particular research question, I dove into the data to see the cause of the unexplained patterns (Figure 11.4).

let us explore the CVR data for interesting insights into the voting that took place in 2020 in Collier County. These insights explain why write-in ballots are problematic, explore the partisan differences in when and how votes are cast, where the Republican Party sees its greatest support (voting pyramids), the demographics of Collier County, and the voting patterns of Collier County over the years.

11.1 The CVR Files

In the 2020 election in Collier County, ballots were cast by voters through the mail, at a central location before election day, or at precincts on election day (November 3, 2020). These ballots were scanned using a specialized computer hardware and software to capture the vote. On the ballots, voters have an option to write in a choice beyond those provided in the ballot. Should the voter decide to do this, a cropped scan of that ballot region was saved instead of which candidate supported.

As an illustrative example, let us assume ballot #55927 apparently wrote in Rvmma Beasley (see scanned image below), which was saved to the CVR file.[1] On the other hand, ballot #55928 voted for Donald Trump. As a result, the record for #55927 consists of a scan of the write-in (as below), whereas the record for #55928 merely records a vote in favor of "REP Donald J. Trump."

- Write-in
  Por Escrito

Because write-in votes have no impact on the Presidential election (see box below), they can be ignored in the analysis—predicated on someone reading through the scanned entries for the name of a major candidate in case someone decided to write "Donald Trump" instead of just filling in the circle next to his name.

---

[1] To be clear, this is *not* an actual scan of a ballot entry (privacy concerns). However, it is based on an actual scan saved to the CVR file. To preserve anonymity, I had a student write-in her favorite literary character. As this is the nicest handwriting she has given me in any class, we should feel privileged. Thanks Nea!!

> **US Electoral Aside**
>
> Because of the vagaries of the US Electoral College, these write-in candidates have no value. This is because the "Presidential election" does not elect the President, it elects Electors to the Electoral College. Since Rvmma Beasley has no registered Electors, this vote has no value beyond that of a protest vote.
>
> For the record, the author of this report believes that this entry is for "Ramona Beasley." However, remember that names are not unique, and there very well could be someone named "Ramona Beasly" or "Ramma Beasley" or even multiple persons named "Ramona Beasley"—across the United States. For whom should this vote count?
>
> This further illustrates the issue of write-in votes for certain elections.

11.1.1 Over- and Undervotes

In addition to the standard votes (fill-in the circle) and the write-in votes, there are two other possible types of votes. An **overvote** occurs when the computer system detects more than one vote for a particular race. An **undervote** occurs when the computer system detects no vote for that race.

An option to reduce the occurrence of over- and undervotes in mail-in ballots is to allow the voter to "cure" the ballot. Not all jurisdictions allow for **ballot curing** of mail-in ballots. Ballot-curing is a process that increases the number of accepted mail-in ballots by allowing the original voter to fix superficial errors in the vote, such as a missing signature, overvoting, and the like. It definitely improves the ability of the voter to ensure their voice is heard.

If ballot curing is not allowed in the jurisdiction, then there are two options to handle over- and undervoting. First, the agents of election supervisor would examine the original ballot to attempt to discern the "will of the voter" and record the vote as such. In the 2000 presidential election in Florida, great effort was expended to determine the will of the voter for each ballot. This led to terms such as "swinging chad," "hanging chad," and "butterfly ballot" entering popular culture.

The second option is to just discard the vote for that particular race (invalidate the vote). It is definitely the easier of the two options. It also removes the person counting from the equation. Furthermore, it allows some statistical testing to occur to determine if invalidating the ballots is independent of whom it was cast for (see Chapter 9). However, there seems to be something distastefully in throwing out someone's vote—especially in a liberal democracy.

11.2 Collier County Background

Much of the following information arises from me doing my due-diligence for the election research. The background offers information from the *Punctum Archimedis*, from the position of the uninterested external viewer interested only in the verifiable truth.

Collier County is in southwestern Florida. As of the 2020 census, the population was 375,752. This is an increase of 16.9% over the 2010 United States Census. The demographic breakdown according to the Collier County Supervisor of Elections is provided in Table 11.1. Note that this information is directly from their website.[Col22b] There are reporting errors in the age category counts. The DEM and REP columns do not add to the purported totals. This error does not exist in the other categories in the table. The errors are minor, however, and have little bearing on the analysis.

Table 11.1 shows additional insight into the "typical" voter in Collier County. They are white (79%) and older than 60 (51%). The voters tend to be more female than male (52% to 46%, respectively). While those females tend to be less Republican than the males (49% to 54%, respectively), this 5 percentage point gender gap is much less than the 18pp in the United States as a whole.[Igi20]

Only the White demographic is Republican (57%). Both the Hispanic (33%) and the Black (7%) demographics are overwhelmingly Democratic. Interestingly, the "Other" group tends to be nonpartisan.

11.2.1 Voting Pyramids

To illustrate the age category information, let us look at two "Voting Pyramids" (Figure 11.1). These show the relationship between age and vote preference.

The left graphic shows the distribution of votes according to age in terms of actual votes. Thus, the "Age 66 and Up" group outvoted any other category (the blue and red bars are wider than any other). The sum of each column of bars is (approximately) 100%. That is, one can see that older voters vote more than younger, regardless of party. This is likely due in large part to the fact that the population of "Age 66 and Up" is greater than any other age

TABLE 11.1

Demographic information regarding residents of Collier County, FL.[Col22b] "DEM" represents those belonging to the Democratic Party; "REP," the Republican Party; and "NPA," to no party (non-partisan).

Demographic	Total Voters	DEM	REP	NPA	Other
White	195,947	36,535	112,112	44,399	2,901
Black	9,396	6,287	625	2,343	141
Hispanic	31,325	10,466	9,662	10,757	440
Other	10,455	2,655	3,710	3,870	220
Male	114,855	21,488	62,281	29,323	1,763
Female	129,720	33,785	63,045	30,976	1,914
Unspecified	2,548	670	783	1,070	25
White Male	92,790	13,839	55,640	21,904	1,407
Black Male	4,314	2,633	390	1,217	74
Hispanic Male	13,692	4,116	4,668	4,721	187
Other Male	4,059	900	1,583	1,481	95
White Female	102,468	22,566	56,152	22,261	1,489
Black Female	4,975	3,581	232	1,095	67
Hispanic Female	17,218	6,200	4,898	5,870	250
Other Female	5,059	1,438	1,763	1,750	108
Sex Unspecified	2,548	670	783	1,070	25
Age 18–25	19,315	5,161	6,696	6,909	549
Age 26–30	11,021	2,893	4,002	3,847	279
Age 31–35	11,376	2,899	4,321	3,893	263
Age 36–40	11,742	2,732	4,829	3,950	231
Age 41–45	12,047	2,848	5,097	3,916	186
Age 46–50	13,120	2,952	6,139	3,817	212
Age 51–55	17,818	3,696	9,330	4,536	256
Age 56–60	22,791	4,611	12,780	5,069	331
Age 61–65	26,187	5,633	14,692	5,504	358
Age 66 and up	101,705	22,517	58,223	19,928	1,037
Totals	247,123	55,943	126,109	61,369	3,702

range (see Table 11.1). Over 25% of all Republican voters in Collier County are age 66 and up.

In contrast, the right graphic shows the distribution of partisanship *by age group*. That is, horizontally the widths add to about 100%. This graphic clearly shows that all age ranges are majority Republican. However, the level

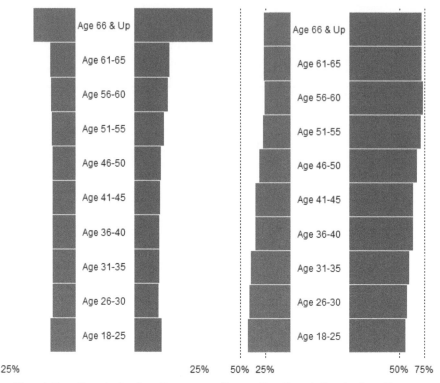

FIGURE 11.1
Voting pyramids to illustrate the level of support in each age group. In the left pyramid, the bars represent the number of votes cast. In the right pyramid, the bars represent the proportion of the vote cast (2020 Presidential election) in each age level.

of that majority decreases as the age of the voter decreases. From age 51 and up, almost 75% of the voters voted Republican in the election.

11.2.2 Brief Electoral History

As Figure 11.2 illustrates, Collier County is historically Republican; it has voted Republican in every presidential election since the author was born.[Lei22]

Since the start of the third millennium, Republican support in Collier County has held rather steady (with a slight and non-significant decline). This does not conflict with the mood of the nation, in which Republican popular vote has tended to hold steady. Not surprisingly, given the national mood, the

Collier County Background 247

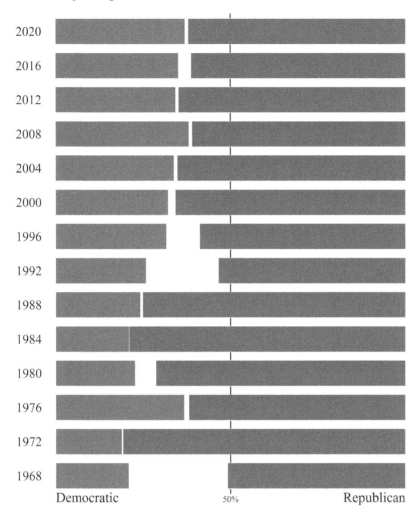

Recorded Vote Proportion

FIGURE 11.2
Voting history of Collier County. Note that the county tends to follow the mood of the country, except for an increase in Republican vote.

two largest Republican wins were in 1972 (Nixon over Humphrey) and 1984 (Reagan over Mondale).

In Collier County, the propensity for voting third-party similarly echoes that of the region and/or the nation. In elections when a popular third-party candidate ran for the presidency, such as Wallace (1968), Anderson (1980), and Perot (1992 and 1996), Collier County tended to vote at similar levels as the region. Additionally, when there is widespread discontent with the two major-party candidates, such as in 2000 and 2016, third-parties captured a not-insignificant amount of the vote.

The proportion of the vote in favor of Donald Trump in 2020 was greater than in 2016. This is most likely due to the shrinking of the third-party vote. It was rather large in 2016, reflecting the widespread dissatisfaction of the major-party candidates. In 2020, it was significantly smaller, reflecting the visceral characteristics of the election. To support this conclusion, note that both major-party candidates increased their vote-share from 2016 to 2020 and that the third-party share dropped dramatically.

11.2.3 Current Voter Statistics

The Supervisor of Elections also provides the current (as of May 6, 2022) distribution of party registration in Collier County.[Col22a]

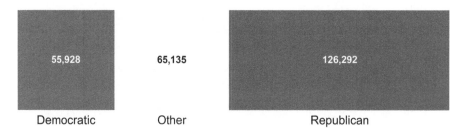

Note that approximately 51% of the voters in the county registered as Republican (23% as Democrats and 26% as other). Thus, under normal circumstances, one would expect an approximate lower bound on Republican vote to be 51% and an approximate upper bound to be 77%.

With that said, every student of (US presidential) elections will contend that no election is held under "normal circumstances." Every election is unique in terms of the personalities of the candidates and the mood of the electorate. Thus, those bounds on Republican vote (51% and 77%) are rather soft in reality.

11.3 Fundamental Question

The author was contacted to investigate a rather simply stated, yet highly complicated to answer, question:

- Is there evidence that there is an issue with the counting of the ballots in Collier County, Florida?

Fundamental Question

This question is simple because it is easily expressed in terms we all understand. This question is complicated because of several issues inherent in voting in a liberal democracy like the United States:

1. votes are not independent in time; the probability that a family votes together (similar political orientation) is not zero
2. votes are not independent in space; precincts are based on geography, and like-minded people tend to live near each other
3. given multiple manners to vote (in-person, by mail, early, day-of-election, etc.), there is no expectation that the populations using these methods are similar in any meaningful way
4. the "correct" probabilistic distribution of votes is unknowable, which means it is not possible to compare the observed distribution to the expected distribution
5. procedures are in place to ensure ballots cast are not invalid, thus one cannot test for differential invalidation
6. the United States utilizes the Australian ballot (a.k.a the "secret ballot"), meaning it is currently not possible to connect a ballot to the person casting it after the vote is recorded

The first five issues I term statistical issues; the last, structural. The statistical issues ensure that meaningful statistical tests regarding the research question cannot be made. Yes, one can perform multiple statistical tests, but they will support the reality of voting distributions being highly particular and unknowable *a priori*.

Some will contend that the distribution of vote counts follows the Benford distribution.[2] However, there is no reason to believe that this is the case, and it is easy to show that such counts do *not* necessarily follow the Benford distribution in clearly free and fair elections (Chapter 8).[For20]

Finally, the last issue ensures that one cannot check that the ballots are not modified between the casting and the counting. This is why so much effort is spent on ensuring the chain of custody is kept.

I provide a solution in the conclusion to this report. Until then, I provide some statistical exploration into the data. This section was the fun part for me. Use this information to gain a deeper understanding of voting in the 2020 presidential election in Collier County and how it does not deviate from expectations.

---

[2] Technically, the Benford distribution is the distribution of leading digits. Thus, it would be more correct to state that the distribution of the leading digits of vote counts always follows the Benford distribution. This is also verifiably untrue.

11.3.1 Statistical Thinking

Before we get to the exploratory analysis of the data, let us spend some time examining the language—and philosophy—of statistical analysis and science. Much of this arises from the philosophical work of Karl Popper and the frequentist work of Ronald Fisher.

Statistical testing starts with a claim about reality (a.k.a., the research hypothesis). This claim divides all possible realities into three: those that are definitely compatible with the claim, those that are definitely not, and those that fit in the grey area between the two.

Once the researcher makes the claim, the statistician attempts to tie the claim to one or more measurables. In some elementary cases, this is easy. For instance, if the claim is that the average height of male students at Florida Gulf Coast University is greater than 70 inches, it is easy to tie the claim to how to measure it (measure male student heights, compare their average to the claimed mean, take into consideration the natural variability in heights).

Such is not always the case.

The current project fits in this latter category. Starting with our knowledge of how votes are cast and how they are counted, we need to theorize methods for corrupting this process. Furthermore, this corruption needs to be detectable.

11.4 Data Preparation and Exploration

Terry sent five CVR files that covered all ballots cast in the 2020 general (including presidential) that were broken into roughly 100,000 records each. This decision was based on the fact that these five files included additional information (votes in other races). This auxiliary information may be useful in understanding relationships amongst the ballots and the recorded votes.

The first step was to combine those five CVR files into one. I copy-pasted from each individual Excel spreadsheet into the final spreadsheet, ensuring that the headers remained the same throughout.

In examining the Excel file, I noticed that the "Marco Island City Council" header covered four columns. I renamed those columns "Marco Island City Council-1," "Marco Island City Council-2," "Marco Island City Council-3," and "Marco Island City Council-4." That was the only change I made to the headers (a.k.a. variables).

I also noticed that the precinct number is usually prefixed with "PCT." However, this was not the case for the two precincts: 434 (8771 records) and

Data Preparation and Exploration 251

479 (3682 records). And so, in the interest of uniformity, I prepended "PCT" to these two labels.

In this data set, there are 417,777 total records documenting 208,866 ballots across 60 precincts (a total of about 54 megabytes of information). A ballot may be split between two records due to some ballots being two pages long. The second page covers votes for two constitutional amendments (No. 5 and No. 6) and a referendum on conservation. Since we are focusing only on the Presidential vote, those secondary records were removed.

11.4.1 Data Observations

The #1 rule to being a statistician is to know the data. This section delves into the data to better understand its features, and to look for outliers and unexpected values. These may indicate typographical errors in the data, whcih need to be fixed.

11.4.1.1 Precinct Distribution

Unsurprisingly, different precincts had different numbers of ballots cast, as seen in the following graphic.

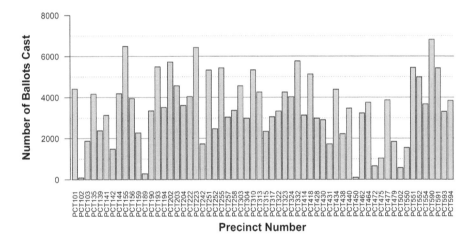

From this graphic it appears as though there are several precincts significantly less populous and several that are more. However, this observation may merely be a result of how the data are presented. Because the precinct number is merely a name for a region of the county, it holds no implicit information. As such, it may be helpful to visualize the distribution of votes cast per precinct in two different manners to detect the "odd" precincts.

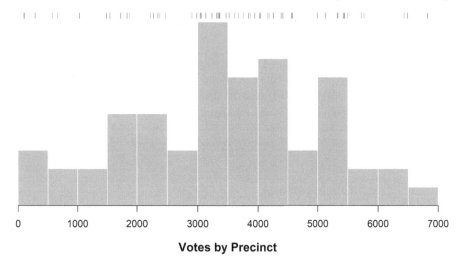

FIGURE 11.3
A histogram of the number of ballots cast per precinct. The lines at the top indicate the actual number of votes cast. The histogram summarizes this information.

The first manner is to calculate the turnout rates for the 60 precincts. This information is not currently available to me. Were it available, one would be able to see which precincts voted at higher (or lower) rates than average. One could use this information to further explore what the under-voting districts had in common and devise manners of encouraging the vote.

The second method is to use a histogram to further illustrate the distribution of vote counts in the precincts. Figure 11.3 does this. Note that the tally marks at the top of the graphic are the actual vote counts in the precincts. The histogram summarizes this information to show an overall trend in the observed data.

The mean number of people voting in a precinct was 6963, with a median of 6984. The minimum and maximums are 182 and 13,655. The distribution of the votes shows no evidence of skew and, surprisingly, is not significantly different from a Normal distribution (i.e., it is entirely reasonable to claim that the votes do arise from a Normal process). Thus, there were no outliers using the typical "MED $\pm$ 1.5 $\times$ IQR" rule.

11.4.2 Vote Timing

The next thing I did was examine *when* the votes were cast. In the interest of upholding our right to a secret ballot, the data do not provide time stamps. However, if we assume that the "Cast Vote Record" number is assigned in

Data Preparation and Exploration 253

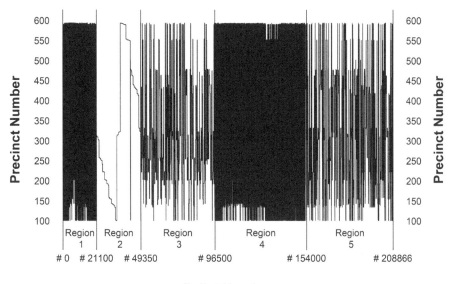

Ballot Number

FIGURE 11.4
A graphic of the precinct number as a function of ballot number being counted. Note that there are five distinct—and three "types" of—regions illustrated.

temporal order, we can graph precinct number against this "quasi-time" variable (Figure 11.4).

Note that there are five separate regions (of three types of patterns) observed. While the exact endpoints of these five regions are not knowable, I approximate them and delineated them using the vertical lines.

The two regions that are essentially black blocks apparently represent time periods when ballots from multiple precincts are processed as they arrive singularly. The first block should represent the early in-person voting; the second, the day-of-election in-person voting. These patterns arise from the fact that the ballots are counted as they arrive, without being ordered by the precinct. The first block consists of ballots from the first until the 21,100th; the second, from the 96,500th until the 154,000th.

The second pattern is mostly white, but with a distinct "slow change" pattern. It occurs from approximately the 21,100th until the 49,350th ballot. In this section, these are the mail-in ballots that were processed by counting each precinct separately. They had been collected for quite some time, and reported after early voting was complete.

This pattern differs from the third pattern in which the mail-in ballots are being processed in much smaller batches, such as in the regions between the 49,350th and the 96,500th ballots and the final region (154,000th until the end). In these two regions, the ballots were being collected and counted quickly. The earlier region represents those ballots received early enough to be processed on or before election day, while the latter region represents those received later.

11.4.2.1 Different Populations

Ultimately, one should fully expect (at least) three of these five regions to be from distinct populations. Back in the 1980s, the mail-in ballots would be overwhelmingly Republican. This was primarily be due to those absentee ballots being cast by the retirees who were on vacation elsewhere or who could not make it to the polls on election day. Since the requirements for utilizing mail-in ballots have been liberalized over the years, the general consensus was that early voting tends to be more Democratic.[LC20]

With that said, President Trump encouraged his supporters to vote early. The effect of this on the usual expectation of early voting was unsure in the run-up to the election. Additionally, the CoViD-19 pandemic and the electoral reactions to it were expected to make the mail-in ballots more Democratic than the in-person voting.[MJOS20] Finally, claiming mail-in voting was an open invitation for fraud, President Trump filed several lawsuits to make drop-off voting more difficult.[3] Thus, taking all of these factors into consideration, we expected the mail-in voting to be Democratic.

The results for Collier County seem to reflect this expectation of early voting being more Democratic than day-of-election voting. Figure 11.5 is a graphic of the same five regions as above. These regions can be combined into four overlapping populations in the electorate: Mail-In (regions 3 and 5), In-Person (regions 1, 2, and 4), Early Voting (regions 1, 3, and 5), and Day-of-Election Voting (regions 2 and 4).

In all five regions, the Republican vote is greater than the Democratic vote (and the minuscule third-party vote). The majority position is indicated by the dashed line in the above graphic. That regions 3 and 5 are more Democratic than the other three regions supports the expectation that vote-by-mail voters tend to be more Democratic than other types of voters.

11.4.3 Vote Relationships

This section examines the relationships between the Presidential vote and the District 19 Representative vote. The correlation between voting for the Party X for President and Party X for Representative is $r = 0.86$ (using Pearson's product-moment correlation), indicating a strong relationship between the two

---

[3]Note that drop-off voting is a version of mail-in voting in which the voter drops the ballot in a specially designed container instead of in a mailbox.

Data Preparation and Exploration 255

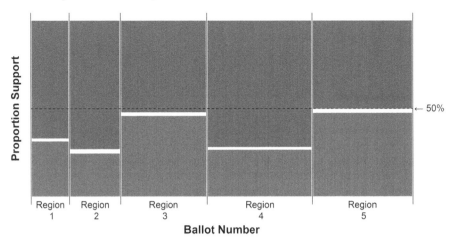

FIGURE 11.5
A graphic of support levels in the five regions. Note that Republicans had a majority in all five regions, but that that majority was most slim in the two mail-in regions.

votes. Over 93% of the voters supported the same party in the Presidential election and the Congressional election. The following graphic illustrates this.

Note that the proportion of those voting for Biden were more likely to vote for Byron Donalds than Trump voters voting for Cindy Lyn Banyai. This would be expected if President Trump had higher levels of negatives than Donalds and if the voters voted for the candidate instead of the party. This observation about candidate negatives matches the national mood in which President Trump had very high negatives, especially as compared to Biden.

Furthermore, it is a well-known fact[4] that Americans tend to "vote for the candidate" instead of the party.

11.5 Conclusion

The purpose of this research was to explore the voting data for the 2020 general election in Collier County, FL. The stated goal was to determine if there is any evidence of the machine changing votes. Since this cannot be done with the available data, the author examined the ballot data for insights that could be used by the sponsors to strengthen democracy in the county (and the country).

While I am here, I would like to thank the Supervisor of Elections of Collier County, Florida, for having this data available for citizens to explore. What she did should be standard practice throughout the United States... and the rest of the democratic world. The data may have raised questions that were eventually answered, but this openness and transparency is what gives people confidence in the electoral apparatus.

Thanks a million!

11.5.1 Next Steps

Ultimately, this is a computer science issue. To determine if the computer systems were biased against one candidate or the other, one would have to fill out a thousand test ballots, and feed them through the system. Comparing the original votes with the reported votes would provide a direct test of the systems used. However, it is quite likely that every machine goes through a tuning process such as this before being used.

---

[4]Whether this really is a fact or just something we keep repeating is unclear. It is certainly repeated enough.

12

System Analysis: Sri Lanka Since 1994

This chapter examines the six Sri Lankan presidential elections between 1994 and 2019 for empirical evidence of persistent electoral unfairness in favor of the government-supported candidates. While regression tests for differential invalidation are the primary methods used, geography will help to illustrate some of the findings—and to provide additional questions.

Republic of Sri Lanka: The Presidential Election of 2015

Among other things, the run-up to the 2015 presidential election in Sri Lanka brought back memories of the consequences of the Rajapaksa-Fonseka 2010 election, which was marked by election-day violence, claims of vote-rigging, and the arrest of a political opponent.[Age15, Cen10, Hum15] It also reminded the region—and the world—of the recent history of Sri Lankan elections. These elections have, without exception, been of a "questionable" nature.

Election-day violence has marred every post-independence presidential election in Sri Lanka.[Gro00, Sri24] Examples of voter-intimidation are frequent; claims of fraud, endemic; and patterns of electoral bias against the ethnic minority Tamil, pervasive.[Age15, Gul13, Hen10, Hum15, Sam89] Election observers have watched the elections. In each election, they have graded Sri Lanka as being less than fully democratic, noting the above intimidation and violence keeping people away from the polls.[Che18, Eur04, Fre17]

However, do such electoral issues of unfairness extend to after the vote is cast? Is there evidence that ballots are treated differently given who they are cast for or given who cast them?

To test such questions of unfairness in the ballot counting, this chapter starts with the necessary, though not sufficient, condition for a fair democratic election we discussed in Chapter 9 and again shows that regression is an appropriate method for testing that necessary condition. Neither ordinary least squares nor weighted least squares should be used, as two requirements are violated by the very nature of this discrete and heteroskedastic data.[MS04]

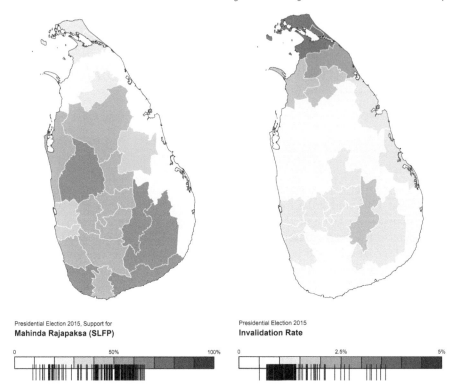

FIGURE 12.1
Maps of the 2015 Sri Lankan presidential election results. The left panel is the vote share for incumbent Mahinda Rajapaksa; the right, the invalidation rate. These data are from the Election Commission of Sri Lanka.[Ele15]

Instead, this chapter uses Beta-Binomial regression, which models a discrete dependent variable that is bounded above and is overdispersed.[Yee15, YW96]

Applying this regression method to the Sri Lankan presidential elections from 1994 until 2019 strongly suggests a consistent unfairness in Sri Lankan presidential elections that always benefits the government-supported candidate. In each of the elections, we find a statistically significant relationship between the invalidation rate and the support for the government candidate. This finding is consistent with differential invalidation. This finding is also consistent with electoral bias against the minority Tamils, who tend to not support the government-supported candidates.

Tamils tend to live in the north and east of Sri Lanka. Figure 12.1 illustrates two things about the 2015 election. First, the left panel shows that the incumbent's area of support was not in the Tamil areas. This is not surprising, as Rajapaksa led the Sri Lankan military in its civil war against the Tamils. Second, the right panel shows that the areas with the highest invalidation rate are those same Tamil areas.

Will we be able to detect differential invalidation over and above that related to the Tamils? Also, was the 2015 an aberration in an otherwise fair electoral process in Sri Lanka?

12.1 Differential Invalidation

Recall from Chapter 9 that many definitions of democracy exist.[DS15, Lij99, Put01, e.g.] In each, a necessary condition for a "free and fair" democratic election is that each person's vote counts the same. While several countries use computers to ensure that all submitted ballots are properly completed before the voter leaves the booth, not all do. In those electoral systems that do not, voters may complete their ballot paper incorrectly. These ballots are then licitly invalidated by the vote counters. In the presence of such invalidation, a necessary condition for a free and fair democratic election is that

> each person's ballot has the same *probability of being counted*, regardless of the voter's demographics or for whom the ballot was cast.

When this is not the case, the election is unfair toward that particular demographic—or toward the supporter of the "wrong" candidate.[EF18]

When ballots are systematically invalidated based on demographics or on candidate preference—a condition called "**differential invalidation**"—then a statistically significant relationship between the invalidation rate and the proportion belonging to that demographic or voting for that candidate in the electoral division may be detectable.[For14] That is, a statistically significant relationship between the invalidation rate and the candidate support rate at the electoral division level is evidence of differential invalidation—and electoral unfairness.

As such, testing for differential invalidation is relatively straightforward, provided that the necessary data are available. Testing for differential invalidation requires the following data at the division level: the number of ballots cast, the number of ballots invalidated, and the number of ballots for the given candidate.[EF18]

12.2 Methods and Data

Since the null hypothesis is that there is no relationship between the two numeric variables (the invalidation rate and the candidate support rate) it may seem completely appropriate to use ordinary least squares (OLS) regression to model the relationship. However, as discussed in Section 9.2.1, two problems arise. The first problem is that OLS requires the dependent variable to be Normally distributed, conditional on the value of the independent variable. With this type of data (counts), this requirement is not met for two reasons: the dependent variable is discrete, and the dependent variable is bounded.

The second problem is that the dependent variable is inherently heteroskedastic.[EF18, KNN04, MS04] Because the dependent variable is bounded below by 0 and above by the turnout, the variance cannot be constant; measurements closer to the upper and lower bounds necessarily have lower variation than those near the middle. Using weighted least squares (WLS) could solve the heteroskedasticity violation, but it does not solve the fact that the data are discrete (counts).

Because of these features of the data, neither OLS nor WLS is appropriate. As in Section 9.2.3, generalized linear models (GLMs) offer a more-general paradigm for handling regression when the dependent variable conditionally follows a non-Normal distribution.[MN00, NW72] As such, one could use Binomial regression via such a generalized linear model. This would allow us to better model the dependent variable. It would also help with the heteroskedasticity. The problem is that the dependent variable does not follow a Binomial distribution. Because similar people live close together, the success probability (base probability of invalidating a ballot) will not be constant across the divisions. This violates one of the requirements of a Binomial distribution. A major consequence of this is that invalidation rates will tend to be overdispersed.[EF18, MS04, Smi02] That is, they do *not* follow a Binomial distribution.

As such, an appropriate regression method must also take this overdispersion into consideration. One can handle overdispersion in a couple of ways. One is to simply allow for the estimation method to estimate the level of overdispersion (quasi-maximum likelihood estimation).[MN00, VR03] This has the advantage of being easily performed using most common statistical programs. The drawback to this solution is that it is relatively restrictive in modeling the distribution of the dependent variable. It only handles dependent variables that are essentially Binomial, but have a higher level of variation than allowed for in a Binomial distribution.

In Section 9.2.4, we discovered an arguably better way to handle overdispersion: We used a different error distribution, one that is more flexible in modeling this overdispersion. One such distribution is the Beta-Binomial distribution.[Ske48, Yee15] This distribution includes the Binomial as a limiting special case. As a result, if there is no overdispersion, then the Beta-Binomial will be as appropriate as the Binomial.

Fitting a dependent variable with a Beta-Binomial distribution is complicated by the fact that this distribution is not a member of the exponential

Methods and Data

class of distributions. Thus, the modeling cannot be performed within the GLM paradigm. A useful extension of the GLM to several additional distribution families takes place using the vector generalized linear model (VGLM) paradigm.[Yee15]

12.2.1 The Beta-Binomial Distribution

The Beta-Binomial distribution is a Binomial distribution in which the success probability parameter π is a random variable that follows a Beta distribution.[Ske48] That is, since the probability mass function of the Binomial distribution is

$$h(x; n, \pi) = \binom{n}{x} \pi^x (1-\pi)^{n-x}$$

and the probability density function of the Beta distribution is

$$g(\pi; \alpha, \beta) = B(\alpha, \beta)\, \pi^{\alpha-1}(1-\pi)^{\beta-1}$$

then the probability mass function of the Beta-Binomial distribution is

$$f(x; n, \alpha, \beta) = \int_0^1 \binom{n}{x} \pi^x (1-\pi)^{n-x}\, B(\alpha, \beta)\pi^{\alpha-1}(1-\pi)^{\beta-1}\, d\pi$$

$$= \binom{n}{x} \frac{B(x+\alpha, n-x+\beta)}{B(\alpha, \beta)}$$

Here, $B(\alpha, \beta) = \frac{\Gamma(\alpha)\Gamma(\beta)}{\Gamma(\alpha+\beta)}$ is the beta function, and $\Gamma(x) = \int_0^\infty t^{x-1} e^t\, dt$ is the gamma function.[1]

Because the Beta-Binomial distribution has two parameters and is not a member of the exponential family of distributions, one cannot use the generalized linear model paradigm. One must use the related vector generalized linear model (VGLM) method.[Yee15] From a practical standpoint, this just requires a different function, `vglm` in lieu of `glm` in the R Statistical Environment.[R 23, Yee10]

12.2.2 Data

As this is ultimately a test of the Sri Lankan government and its claim of a "free and fair" election, the data are the official counts as provided by the Sri Lankan Election Commission website.[Ele94, Ele99, Ele05, Ele10, Ele15,

---

[1] For the mathematicians amongst us, note that as $\alpha \to \infty$ and $\beta \to \infty$, the Beta-Binomial distribution converges in distribution to the Binomial distribution. I leave this as a fun exercise for you.

Ele19] Sri Lanka has nine provinces, 25 districts, and a total of $n = 160$ electoral divisions.[Ele18b]

Due to data scarcity brought about by the civil war (1983–2004), the Tamil-dominated districts in the north (Jaffna and Vanni) were removed from consideration in 1994, 1999, and 2005. For the sake of consistency, they were also removed for the elections of 2010 and 2015.[2] Note that removing the northern electoral divisions means any evidence of differential invalidation is independent of the Tamil question.

All of these deletions were done to make the model statistically more sound. From the standpoint of the substantive conclusions, however, these deletions tend to not be practically significant. Performing the analyses with the entire data set changed the substantive results only in the 2015 election, where the differential invalidation only becomes statistically significant with the addition of Jaffna and Vanni districts.

The dependent variable in the model is the count of invalidated ballots. The independent variable is the proportion of the votes cast in favor of the government candidate. Over the course of the five elections, the invalidation rate at the electoral division level ranged from 0.48% (Galgamuwa in 2010) to 5.94% (Haputale in 1994), with a median of 1.27%. The support rate for the government candidate at the division level ranged from 3.6% (Rajapaksa in Padiruppu in 2005) to 94.2% (Kumaratunga in Mahiyanganaya in 1994).

12.3 Results by Election

The numeric results of the analyses are provided in Table 12.1. The first column provides the election year and the main candidates, with the government-supported candidate listed first. The second column gives the estimate, and the final column the endpoints of the typical 95% confidence interval for the candidate support effect.

12.3.1 The 1994 Election

The November 1994 election was set during the Sri Lankan civil war. In 1993, the Tamil Tigers (LTTE) successfully assassinated President Ranasinghe Premadasa of the United National Party (UNP). Prime Minister Dingiri Banda Wijetunga succeeded him, but decided to not run in the upcoming

---

[2]Also removed are the "Postal" votes and "Displaced" votes, as they cannot be assigned to a division from the information provided by the Sri Lankan government.

TABLE 12.1
Regression results for the six elections. The numbers in the second column are the estimated effect levels. Those in the last column represent the endpoints of the central 95% confidence interval for the support effect parameter. Negative candidate support intervals indicate statistically significant evidence of unfairness in favor of that candidate. The second set of numbers for the 2015 and 2019 elections are for the model including the Jaffna and Vanni districts.

Election	Estimated Support Effect	
1994 (Kumaratunga v. Dissanayake)	−1.90	(−2.72, −1.09)
1999 (Kumaratunga v. Wickremesinghe)	−1.07	(−1.63, −0.52)
2005 (Rajapaksa v. Wickremesinghe)	−1.08	(−1.36, −0.81)
2010 (Rajapaksa v. Fonseka)	−2.28	(−2.60, −1.96)
2015 (Rajapaksa v. Sirisena)	−0.28	(−0.61, 0.05)
	−1.08	(−1.42, −0.73)
2019 (Rajapaksa v. Premadasa)	−0.67	(−0.95, −0.38)
	−1.42	(−1.64, −1.20)

election. Because of this decision, the UNP named Gamini Dissanayake as their candidate.[Hen10]

Dissanayake's challenger was Prime Minister Chandrika Kumaratunga of the People's Alliance (PA). Kumaratunga became prime minister when her party took control of the Parliament in the August 1994 parliamentary elections. Because her party controlled parliament, she is considered the "government-supported" candidate in this election. Kumaratunga received 62% of the votes counted in the November presidential election.[US 04]

To perform this analysis, we load the 1994 election file for Sri Lanka and load the VGAM package. Then, to use the Beta-Binomial distribution to model the invalidation rate, we use this code

```
tot = dt$Total
inv = dt$Rejected
val = tot-inv
px  = dt$PA/val   ## People's Alliance candidate support

depVar = cbind(inv,val)
modBB94a = vglm(depVar~px, family=betabinomial)
summary(modBB94a)
confint(modBB94a)
```

The first block of code captures the number of invalid ballots, total ballots, and votes for the People's Alliance candidate, as reported by the government (with Jaffna and Vanni districts removed). The bottom block of code performs the regression, produces the regression table, and estimates the 95% confidence intervals for the estimates.

The heavily abbreviated output from this code is

```
Coefficients:
            Estimate Std. Error z value Pr(>|z|)
px           -1.9020     0.4167  -4.565   5e-06 ***
```

and

```
              2.5 %     97.5 %
px         -2.718664 -1.085398
```

This output tells us that the estimated effect of Kumaratunga support on the logit of the invalidation rate is -1.9020 with a 95% confidence interval from -2.718664 to -1.085398. This effect is statistically distinct from 0 because the p-value is so small (about 0.000005).

In other words, differential invalidation was detected in this election that favored the government candidate. The point estimate of -1.9020 indicates that for every 10-percentage point increase in support for Kumaratunga in the division, the probability of a ballot being invalidated decreases by about 17% ($= 1 - e^{-1.0920 \times 0.10}$). That is, ballots were more likely to be counted in divisions that supported the government candidate than in those divisions that did not.

Figure 12.2 illustrates this conclusion: Divisions with higher levels of support for Kumaratunga had a lower invalidation rate—i.e., a greater proportion of their votes counted. This is *prima facie* evidence of unfairness in this election. Since the Tamil-dominant areas are not a part of this regression, this is evidence that the differential invalidation specifically helped the government-supported candidate.

Note that this is just a test and is not for retrieving the true number of ballots cast for the candidates, nor is it an argument that the challenger would have won had the differential invalidation not taken place. To perform those analyses, one would have to determine what the "real" invalidation rate is for each division. This is currently beyond the scope of what we are able to accomplish—currently.[3] The purpose of this analysis is merely to look for evidence of differential invalidation, not to determine if that differential invalidation swung the election.

---

[3]There is some exciting research taking place at Baylor University that applies to this particular problem. Nelson et al. are looking at Bayesian methods for retrieving correct counts from error-prone data.[NSCS18] While they are specifically looking at crime rates, their choice of framing this in terms of count data is exciting.

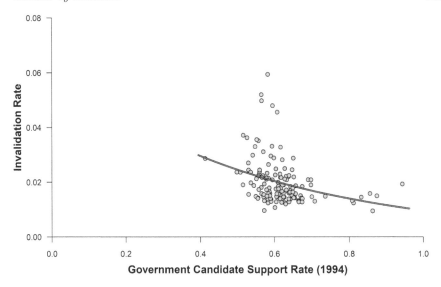

FIGURE 12.2
A plot of the invalidation rate against the support for the government candidate in the 1994 election. The dots represent the electoral divisions. The negative slope to the regression curve indicates differential invalidation helped Kumaratunga in the election.

12.3.2 The 1999 Election

The December 1999 election saw Kumaratunga defend her presidency against Ranil Wickremesinghe of the United National Party (UNP). Kumaratunga's People's Alliance remained in control of the parliament, making her the government-supported candidate. She won a second term with 51% of the votes counted.[Hen10, Sha03, US 04]

A similar analysis produced the following abbreviated output

```
Coefficients:
          Estimate  Std. Error  z value  Pr(>|z|)
px         -1.0733     0.2837   -3.784   0.000154 ***
```

and

```
           2.5 %      97.5 %
px       -1.629266  -0.5173614
```

This output shows that there is significant evidence of differential invalidation in this election, too. On average, for every 10-percentage point increase in

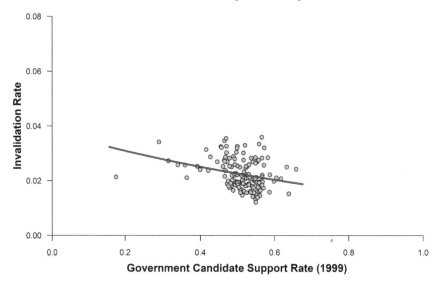

FIGURE 12.3
A plot of the invalidation rate against the support for the government candidate in the 1999 election. The dots represent the electoral divisions. The negative slope to the regression curve indicates differential invalidation helped Kumaratunga in the election.

support for Kumaratunga, the invalidation rate drops by approximately 10% ($= 1 - e^{-1.0733 \times 0.10}$). Figure 12.3 illustrates this effect.

Again, there is strong evidence of unfairness in this election, and that unfairness helped the government candidate. Again, since the northern electoral divisions are not a part of the analysis, this unfairness is over and above any unfairness in the electoral system against the Tamils.

12.3.3 The 2005 Election

Parliament underwent some significant changes between the 1999 and 2005 elections. The biggest change was that President Kumaratunga's People's Alliance party lost its majority in the parliament. In fact, from 2001 until 2004, Wickremesinghe's United National Party (UNP) held the majority and the premiership.

To help ensure that her coalition controlled the parliament, Kumaratunga merged the People's Liberation Front and the People's Alliance to form the United People's Freedom Alliance (UPFA). The move was successful, as the 2004 parliamentary election saw the UPFA win a majority of the seats.[Hen10] This placed her party in power in time for the 2005 presidential election.

However, Chandrika Kumaratunga did not run for a third term in 2005. In her stead, Prime Minister Mahinda Rajapaksa ran against Wickremesinghe of the United National Party (UNP). Thus, because his UPFA party controlled the parliament, Mahinda Rajapaksa was the government-supported candidate in the 2005 election.[Hen10, Uya10b]

The analysis of the 2005 presidential election produced this abbreviated output:

```
Coefficients:
            Estimate  Std. Error  z value  Pr(>|z|)
px          -1.08495    0.14070   -7.711   1.25e-14 ***
```

and

```
              2.5 %      97.5 %
px         -1.360718  -0.8091844
```

These results show significant evidence of differential invalidation in favor of Rajapaksa, the government-supported candidate. Given that there is no differential invalidation, the probability of observing results this extreme are one in 10^{14}—rather rare. Furthermore, for each 10-point increase in support for Rajapaksa, the probability of a ballot being invalidated decreases by an average of 10%. Figure 12.4 illustrates this evidence of unfairness in the 2005 election.

There are two important things to note about this election beyond the differential invalidation. First, note that the invalidation rate tended to be much lower in this election than in the previous two. This may (or may not) indicate that the 1994 and 1999 elections had inflated invalidation rates arising from Kumaratunga supporters happily invalidating the votes originally destined for her opponents.

Secondly, note that the slope of the regression curve is also much shallower here than in the 1994 and 1999 elections. This, and the previous observation, strongly suggests that this election and the previous two did not follow the same "rules of invalidation."

12.3.4 The 2010 Election

The Sri Lankan civil war officially ended in May 2009.[Wij09] The January 2010 election had President Rajapaksa challenged by Field Marshal Gardihewa Sarath Chandralal Fonseka. While the two worked well together in the closing years of the war, the two disagreed on several issues, including the future of Sinhalese-Tamil relations in Sri Lanka.[Hen10, Uya10a]

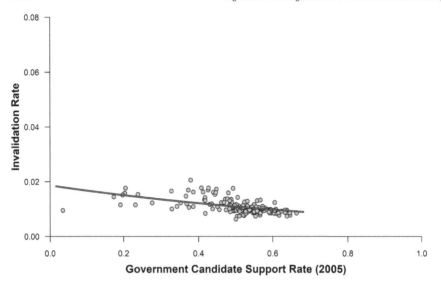

FIGURE 12.4

A plot of the invalidation rate against the support for the government candidate in the 2005 election. The dots represent the electoral divisions. The negative slope to the regression curve indicates differential invalidation helped Rajapaksa in the election.

As no parliamentary elections had been held since 2004, Rajapaksa's United People's Freedom Alliance held the premiership in the name of Ratnasiri Wickremanayake. As such, Rajapaksa is the government-supported candidate. Again, the government-supported candidate won the election. This time, Rajapaksa received 57% of the ballots counted to Fonseka's 40%.

This is of particular note because the international news services who had done some polling thought that the election would be so close that the winner would not be known for some time.[BBC10c] Furthermore, while the ballots were being counted, Rajapaksa sent the army to "secure" the Cinnamon Lakeside hotel, in which Fonseka and his advisors were staying.[BBC10b] This eventually led to Fonseka's arrest and convictions for irregularities in army procurement and for war crimes.[BBC11b]

Table 12.1 shows that there was significant evidence of differential invalidation. For every 10-point increase in support for Rajapaksa, the invalidation rate in the division decreased by an average of 20%. Figure 12.5 illustrates this evidence of unfairness in the 2010 election.

This election shows the strongest differential invalidation of any being investigated. It is also the election with the gravest results. Election day was particularly violent and Fonseka, the opposition candidate, was arrested just

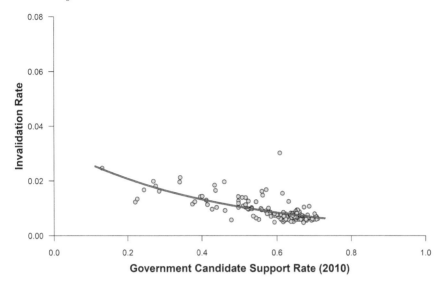

FIGURE 12.5
A plot of the invalidation rate against the support for the government candidate in the 2010 election. The dots represent the electoral divisions. The negative slope to the regression curve indicates differential invalidation helped Rajapaksa in the election.

a couple days after the election. This memory weighed heavily on the minds of Sri Lankans as they went to the polls in 2015.

12.3.5 The 2015 Election

The April 2010 elections to the 14th Parliament of Sri Lanka increased Rajapaksa's power in Sri Lanka. His United People's Freedom Alliance (UPFA) held an absolute majority in the Parliament of Sri Lanka. This allowed his UPFA to amend the constitution, increasing the power of the president—and eliminating term limits.[BBC10a] Because of the large amount of power his party held, Rajapaksa felt certain he would win re-election. However, alliances changed rather swiftly after Rajapaksa announced the early elections and his desire to be reelected for a third term.

His opponent was Maithripala Sirisena, Rajapaksa's Minister of Health and a former member of Rajapaksa's UPFA coalition. The opposition United National Party (UNP) selected Sirisena as their candidate. Almost immediately, many of Rajapaksa's allies switched to support Sirisena.[Tam14]

Table 12.1 shows slight evidence of differential invalidation ($p = 0.0929$). However, when the Tamil-majority districts of Jaffna and Vanni are included,

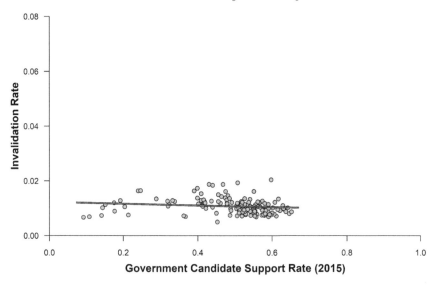

FIGURE 12.6
A plot of the invalidation rate against the support for the government candidate in the 2015 election. The dots represent the electoral divisions. The (slight) negative slope to the regression curve indicates differential invalidation helped Rajapaksa in the election.

the evidence and the effect size both become much stronger ($p < 0.0001$). Figure 12.6 illustrates this evidence. On average, for each 10-point increase for Rajapaksa, the invalidation rate drops by an average of 10%.

From a statistical standpoint, this is the most interesting of the elections. Without the Tamil regions, there is no real evidence of differential invalidation in favor of Rajapaksa—perhaps because he lost the election. However, with the inclusion of the Tamil areas, there is evidence. This strongly suggests that the election was biased against the Tamils. Whether this is due to conscious biasing of the counters against the ethnic-minority Tamils *or* due to the electoral rules that hampered the Tamils and their votes from counting, one cannot directly tell. It does help illustrate the invalidation map for the 2015 election, however (Figure 12.1). Those areas that supported Rajapaksa were not the Tamil-dominated areas in the north. Similarly, those areas that had higher-than-average invalidation rates *were* those Tamil-dominated areas in the north.

12.3.6 The 2019 Election

While the Sri Lankan constitution allowed President Maithripala Sirisena to serve a second term, he announced on the day of his inauguration that he would serve but one term.[SBS15] This left the election without an incumbent. To fill that vacuum, the two major parties selected Sajith Premadasa (UNP) and Gotabaya Rajapaksa (SLPP). Premadasa is the former Deputy Minister of Health (2001–2004 under Kumaratunga) and the son of assassinated former president Ranasinghe Premadasa. Rajapaksa is a former Secretary to the Ministry of Defence and Urban Development (2005–2015 under Rajapaksa) and the brother of former president Mahinda Rajapaksa.

Because of internal power struggles in the previous two main parties, as well as the formation of the Sri Lanka People's Front (SLPP) party out of several SLFP members who supported former president Mahinda Rajapaksa, it is rather difficult to determine who is the "government-supported" candidate. However, after a devastating loss in the 2018 local elections, Sirisena's Sri Lankan Freedom Party (SLFP) decided to support the SLPP candidate, Gotabaya Rajapaksa.[New19] This, along with the fact that his brother is a former president and big figure in Sri Lankan politics, arguably makes him the government-supported candidate.

Table 12.1 shows strong evidence of differential invalidation ($p < 0.0001$). In fact, when the Tamil-majority districts of Jaffna and Vanni are included, the evidence and the effect size both become *much* stronger. Figure 12.7 il-

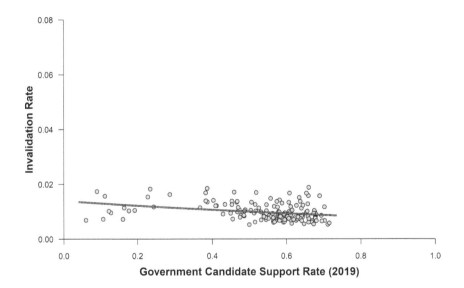

FIGURE 12.7
A plot of the invalidation rate against the support for the government candidate in the 2019 election. The dots represent the electoral divisions. The negative slope to the regression curve indicates differential invalidation helped Rajapaksa in the election.

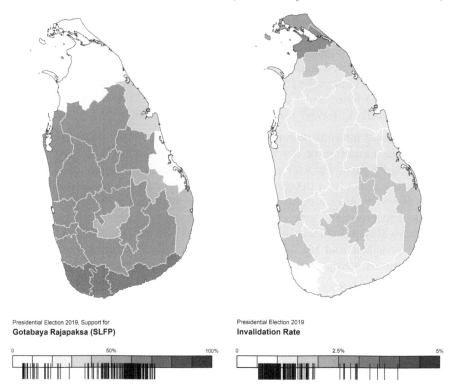

FIGURE 12.8
Maps of the 2019 Sri Lankan presidential election results. The left panel is the vote share for Gotabaya Rajapaksa; the right, the invalidation rate. These data are from Election Commission of Sri Lanka.[Ele19]

lustrates this evidence. On average, for each 10-point increase for Rajapaksa, the invalidation rate drops by an average of 6.4%.

Finally, note that Figure 12.8 illustrates the effects of differential invalidation on this election. The left panel shows the support rate for Rajapaksa; the right, the invalidation rate. Rajapaksa received very little support from the Sri Lankan Tamil areas (north and east). Furthermore, he received less-than-average support from the areas with many Indian Tamils (south of center in the map). These Tamil areas also tended to have higher-than-average levels of invalidation.

Is this due to the counters purposely invalidating many votes for Rajapaksa? Is this due to the electoral system being biased against those who do not speak Sinhalese? With this analysis, one cannot tell. However, giving the government the benefit of the doubt suggests that the government should undertake a study to determine if the electoral system is at fault.

Discussion

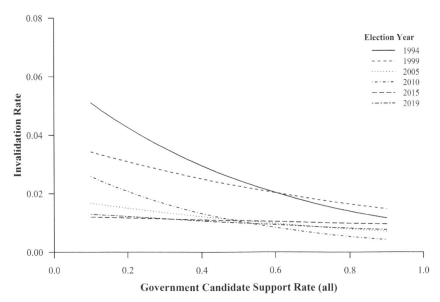

FIGURE 12.9
A plot of the six regression curves. Note that all six have a negative slope, indicating differential invalidation that helped the government-supported candidate in each and every election. The probability of all six helping the government-supported candidate by accident is infinitesimal.

12.4 Discussion

Figure 12.9 provides all of the regression curves on one graphic. Note that all slopes are negative. This indicates that divisions supporting the government-supported candidates tended to also have a lower proportion of their votes declared invalid. This is consistent with systematic differential invalidation: the "rules" for invalidation seem to differ between those votes in favor of the government candidate and those against. If the rules were the same, then the slope (candidate support effect) would be near zero.

The fact that the slope is negative in *all of the elections* is very striking. Were there no inherent bias toward the government-supported candidate, then we would expect some slopes to be positive and some to be negative. Here, *all* are negative. This is extremely strong evidence either of a problem with the electoral laws/system or with how the votes are counted.[4]

---

[4] Actually, this could also be evidence of persistent ballot-box stuffing (see footnote on page 209). However, it is rather difficult to explain how ballot boxes were stuffed for 25 years without some direct evidence in the form of photographs or testimony.

While the regression slopes are all negative, indicating unfairness in the elections in favor of the government-supported candidate, the 2015 election shows the least evidence of differential invalidation. This may be due to a few reasons. First, the "government-supported" candidate lost a lot of government support in the days before the election. There was a relatively large exodus of UPFA members to the opposition bloc.[Tam14] Thus, Rajapaksa may not have been as government-supported as one would expect.

Second, the differential invalidation may have been mitigated by the fact that the government candidate lost. That is, while differential invalidation did exist in the election, its effect were lessened by the fact Rajapaksa-supporters did not control most divisions.

Third, it is interesting that whether or not the two Tamil-majority districts are included largely determines our conclusion in this election. When they are included, the differential invalidation becomes highly significant ($p < 0.0001$). Since these two districts supported challenger Sirisena (Jaffna 74% to 22% and Vanni 78% to 19%), this means the Jaffna and Vanni districts experienced higher-than-expected invalidation.

12.5 Conclusion

Researchers frequently define "free and fair" democratic elections. While the definitions tend to vary on the finer points, they agree that fair democratic elections require that the probability of a person's ballot counting must be independent of whom the vote was cast for. When there is a relationship between the invalidation rate and the support rate for the candidate, the election exhibits differential invalidation, which is a result of one type of unfairness. Testing for this relationship is as easy as using regression—albeit one that takes into consideration the discrete nature of the dependent variable, as well as its inherent heteroskedasticity and overdispersion. If a relationship is detected, then there is significant evidence of unfairness in the election.

This research examined the six Sri Lankan presidential elections from 1994 to 2019 for evidence of persistent unfairness. In each of the elections, significant differential invalidation is evident. This suggests changes need to be made to the electoral structure in Sri Lanka to help ensure free and fair democratic elections.

Sirisena made some changes to the system in 2015. The Sri Lankan Department of Elections oversaw elections from 1955 until it was replaced by the Election Commission of Sri Lanka on November 17, 2015—after the 2015 Presidential election.[Ele18a] This change seems to have not resulted in eliminating the differential invalidation; the 2019 election also showed that the government-supported candidate benefited from it.

Note that the slope of the 2019 election is rather small. Thus, it may be that the changes wrought by the Election Commission of Sri Lanka are there; they just did not appear in that particular election. Further election may show the differential invalidation going against the government-supported candidate. It may even show a lack of differential invalidation. If either of these scenarios comes to pass, then we will have evidence that the Election Commission has succeeded in making Sri Lankan elections more free and fair.

Until then, we can only hope.

12.6 A Sri Lankan Postscript

In 2020, Sri Lanka underwent a series of significant events. First, the country faced a protracted economic crisis, which was arguably exacerbated by the National Thowheeth Jama'ath (NTJ) terrorist bombings on Easter Sunday, 2019.[Al 19] Although this attack worsened Sri Lanka's pre-existing economic challenges, it did not singlehandedly cause them.

Furthermore, in July 2022, President Gotabaya Rajapaksa resigned from his presidency while in self-imposed exile in Singapore, following widespread protests against his handling of the economic crisis. A few days prior to his resignation, Rajapaksa attempted to flee Sri Lanka but was unable to do so because of passport issues (the immigration officials refused to process them). The next morning, he and his entourage left for the Maldives on a Sri Lankan Air Force plane. Later, he flew to Singapore.[IAR22]

Following Rajapaksa's resignation, Ranil Wickremesinghe became Acting President and was subsequently elected as President by the Parliament on July 20. This was in accord with the Presidential Elections (Special Provisions) Act of 1981.[MK22]

Despite these leadership changes, Sri Lanka's economic situation remains challenging and unresolved. As a result, Wickremesinghe is badly trailing Anura Kumara Dissanayaka, the leader of the National People's Power (NPP), in opinion polls. For instance, the November 2023 poll held by the Institute for Health Policy had Dissanayaka ahead of Wickremesinghe 51% to 13%.[RE23]

Who will win the 2024 Sri Lankan presidential election? As the Danish maxim goes, "It is very hard to predict, especially the future."

13

An Afterword: Biases in Polling

This final chapter takes a break from the experimentation and the mathematics and discusses several enduring issues in polling. These include the problems with the population and with sampling. They also include properly representing the results.

Make no mistake about it: Representing the results perfectly is all but impossible. Readers are looking for different things from the article. And, each of the different things has a different need from the presentation. The key is clarity.

This final chapter concludes with my own diatribe extolling the virtues of analysts retaining ownership of their analyses. This means they are partially responsible for ensuring that the results are conveyed to the end-user properly.

United States of America: The Presidential Election of 2020
Of course I had to end the book with the 2020 presidential election in the United States. In its aftermath, one of two main candidates claimed that there was electoral fraud taking place. However, when given the chance to present his evidence, neither he nor his many allies were able to present any evidence that held up in a court of law. This was quite the blow to them.

Part of the reason, I believe, that there was such widespread belief in electoral fraud was a basic misunderstanding of what polls were telling—and *could be* telling—about the election. Polls in many states showed the president ahead of his challenger. However, there was a great amount of variation in those polls. Those that showed the president ahead either had a large confidence interval that contained the "losing to his opponent" or were balanced out by other polls showing his opponent ahead.

These dueling polls confused the electorate a great deal. This confusion was compounded by the fact that different news organizations tended to pay for their own polls from familiar polling firms. From the stratified sampling chapter (Chapter 4), we remember that all polling firms must estimate the target population— those who will vote. These will all be incorrect in some way, but in different directions and in different amounts. Thus, if the go-to polling firm for a news network overestimated the Republican vote, then the listeners to that news stations would expect the Republican candidate to do much better than he did. Similarly, if the news network's polling firm overestimated the Democratic vote, then their estimates would tend to underestimate the president's support in the population.

And so, when election day arrived, we were able to see how well the polling firms did with estimating the target population. Some overestimated the Republican vote. Some overestimated the Democratic vote. None were perfect in all 50+1 elections.[1]

And so, because people tend to get their news from a single source, and that single source tends to obtain poll estimates from a single source, and *that* single source weighted the same throughout the election cycle, Americans tended to get a skewed view of the election.

Figure 13.1 shows the electoral performance of Donald Trump in the 2016 and 2020 elections for states that he won. The interesting map, in my opinion, is the bottom one. That map shows the change in vote proportion from 2016 to 2020 for Donald Trump. Note that he only increased his vote share in 7 of the states: California, the District of Columbia, Florida, Hawai'i, Illinois, Nevada, and New York.

Of these seven states, Trump won only Florida in 2020 (and in 2016). This means that he lost vote proportions in all states he won in 2016 *except* for one. This surprising result, I believe, also led to his voters believing there was electoral fraud in the 2020 election.

13.1 The Sources of Bias and Variance

I have tried to weave three threads throughout this book. The first is that statistics is very useful in understanding elections around us, from the polls to the analysis. It almost seems as though elections speak using the language of probability and statistics.

---

[1] In the United States, each state elects people to serve on the Electoral College. It is these Electors who elect the president. Thus, there are really 51 elections taking place— one for each of the 50 states, plus the District of Columbia. Throughout this chapter, I will use the term "50+1 states" when referring to the 50 states plus the District.

Also note that Maine and Nebraska can split their electors. All other states are winner-take-all. In 2020, both states split, with Maine electing one Republican Elector; and Nebraska, one Democratic.

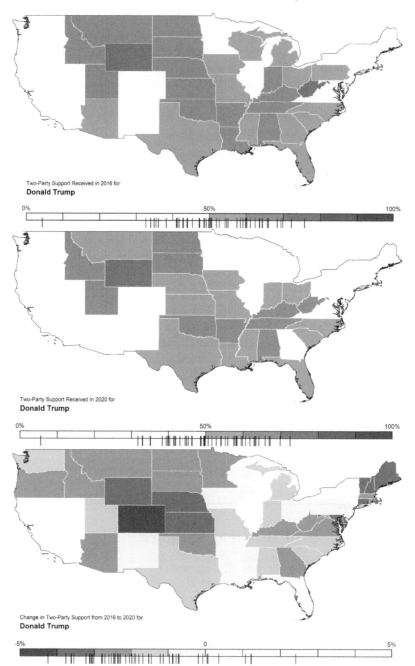

FIGURE 13.1
Result maps of the 2016 (top) and 2020 (middle) presidential election in the United States. The bottom map is the difference in Trump support between the two elections for the 43 contiguous states he did worse in. These data are from the US Federal Election Commission.[Wi24a][Wi24b][Fed22]

13.1.1 Variance: Instability in the Estimates

Second, there is a source of variability in poll results that can *never* be eliminated (Section 2.2.1). The sample is random. Anything measured on that sample, such as candidate support, is also random. The distribution underlying all polling is the Binomial distribution. The variability that is simply due to random sampling is $\pi(1-\pi)/n$, where π is the issue support in the population (what we are trying to estimate) and n is the size of the sample.

While this variability does decline as the sample size increases, it never goes to 0. Furthermore, reducing it requires a larger sample size, and collecting data is expensive. To obtain a margin of error of just 5 percentage points, one needs a sample size of about 400 (Section 2.2.5). To reduce that margin to 4pp requires 625—more than a 50% increase in data collection. A margin of error of 2pp requires 2500. In general, to cut the margin of error in half, one needs to collect **four times as much data**, which would increase the cost dramatically.

This variability cannot go away. It is a part of the very fabric of the universe.

13.1.2 Bias: Systematically Wrong

The third thread I have tried to weave throughout the book is that it is of paramount importance to understand the target population. It is also impossible to know it perfectly. At best, we can get close to it.

Mind the Population Gap

Recall from Section 1.2.1 that there are two populations at work when sampling. The population from which the sample is drawn is the sampled population. The population we are drawing conclusions about is the target population (or "population of interest"). In terms of polling, the target population is the group of voters who *actually* vote in the election. The sampled population is the group of people who can be contacted by the call center *and* who the polling firm believes are a part of the target population. The two populations are not the same.

This is problematic.

It is also unavoidable.

Polling firms tend to use stratified sampling to mitigate this problem. However, that raises additional problems, especially those covered in Chapter 4. It is these "fixes" that both improve the problem and introduce bias.

Missing *Not* at Random

There is another source of bias in polling: not all parts of the population are equally accessible. The landlines are getting fewer and fewer in the OECD countries. More and more households are becoming cell-phone-only households. Caller ID allows individuals to screen their calls, thus letting the unrecognized numbers go to voicemail. All of this undercuts the basic assumption that all members of the sampled population are equally accessible.

These trends have only accelerated over the past generation. It is to the point that the author is one of those people who has no landline and will not answer a phone call from an unrecognized number. I am a part of the problem.

As long as there is no correlation (relationship) between the likelihood of taking part in a poll and the political orientation (that is, they are "missing completely at random; MCAR), the above only increases the variability in the poll results. However, they *are* related. As a result, those left out of the poll are "missing not at random" (MNAR).[LR02]

For those who study sampling, MNAR is the worst, as you imagine. There is little one can do when the data are MNAR. Fixing the problem (avoiding the bias) requires an intimate understanding of the population. Unfortunately, that is something we do not know. For instance, if 30% of the population cannot be contacted for a poll, we need to *guess* how they will vote.

Bots: The Un-Respondents

Some sampling schemes, especially those fully or partially utilizing online polls, are susceptible to fraudulent responses. This goes beyond people lying on surveys because "It's no one's business how I will vote" (and I am not convinced this group is a significant factor).

Recently, on the AAPOR Listserv, this very issue was raised by a small academic polling shop. The group used online surveys for their research. Unfortunately, their surveys were being hit by bots. This is, to say the least, quite problematic—especially for surveys that need to ensure the respondents are kept anonymous.

Thankfully, the Listserv members suggested several solutions. Unfortunately, the vast majority of these solutions all had at least one of two problems. The first possible problem is that the solution made it more expensive for the polling shop to operate. In this time of reduced budgets, this cuts back on the amount of good research that can be done. The second possible problem is that the solution made it more difficult for real respondents to respond. Raising the hurdle reduces the sample size that is collected.

The presence of bots causes problems.

13.2 A Concluding Diatribe: The Blame for 2020

The 2020 presidential election in the United States shows the problems that arise from not truly conveying what polls say—and what they do not. Make no mistake about it: The 2020 election was a long time coming. Over the past generation (at least), the understanding of polls has declined dramatically. While I am based in the United States, it appears to me that this phenomenon is world-wide.

I blame three groups for this. First, I blame the **polling firms** and analysts. The data and the analysis is ours. We need to take rightful ownership of both and steward them to the very end. We can no longer just take the money and send the analysis to the customer (news station, media, etc.). We need to ensure that those results are properly presented to the people. This includes all of the uncertainty and caveats that are second nature to us.

While it does feel like it undercuts our message, ensuring that the uncertainty is understood ensures that we are treated like the professional scientists we are.

The second group is the **corporate media**. A business model that is based on viewership is destined to cater to the audience. There is pressure to report news that fits the viewers' mindset. This affects the choice of polling firm. It also affects the interpretation of those polling results.

The third group is the **citizenry**. We need to be better informed about statistics and polls—if we care about them. This comes from our schooling, yes, but it also comes from the news we consume—*especially* after we leave formal education.

The news media *educates us*. Thus, it is even more important for the news to tell the whole truth about the polls. But, it all comes back full-circle to the polling researchers. We cannot give up ownership of our analyses. It is our duty.

A Final Thank You. Together, we started this journey, took that first step, over 280 pages ago. I hope it has been well worth your while and that you reached your goal of understanding elections through statistics.

∼ Ole J. Forsberg
   Galesburg, Illinois, USA
   March 26, 2024

A

A Brief Introduction to R

This book relies on using the R Statistical Environment for calculation, estimation, and experimentation. There are excellent reasons for using R, as discussed in this appendix. However, as with all software, there is a learning curve to experience. This appendix seeks to provide you a solid foundation in coding using R.

Often, there is much consternation among aspiring applied statisticians as to why they have to learn *yet another* statistical package. Can't those dad-gum statisticians make up their minds? Are they trying to drive us crazy? Are they getting kick-backs from the statistical software salesmen?

The answer, more than likely, is that the specific author truly believes that *this* statistical package is the best available (by whatever measure they use). I am no different. I truly believe that, overall, the R Statistical Environment is the best statistical package available for the following four reasons:

- It is free (both free pizza and free speech).
- It is flexible.
- It is powerful.
- It matches how science *should* be done.

Let us take these four reasons in the above order and explain what I mean by each.

First, it is free. The "R Project for Statistical Computing" is a not-for-profit foundation created for the sole purpose of creating a piece of software that encourages scientific innovations in any field that uses statistics.[1] The cost of the software is zero. That means students (and professors and businesses and journalists) do not have to pay fees (license fees, purchase fees, or annual rental fees) to use the software. Furthermore, because you are able to load it on a USB drive, you do not have to wander around searching for a computer sporting R like you do with most other statistical packages. It is also free in

---

[1] The URL for the R Project is https://www.r-project.org/.

the sense that you are free to modify the code to make it better. This latter part is what allows R to become better, stronger, and faster as time passes.

Second and third, it is flexible and powerful. The base distribution of R contains only those parts of R that are universally useful. Thus, it is small and fast to download and begin. However, there are assorted packages for just about any statistical analysis. And, for those methods that are not currently supported by base R, you are free to create your own functions and packages for everyone to use. Furthermore, it is a scripting language. As such, you can program it to do the same tasks repeatedly (with slight modifications) so that robust analyses can be done.

Finally, it matches how science *should* be done. R offers a definite separation between the data and the analysis. It also offers a way of keeping track of your analysis as you *do* the analysis. The former allows you to keep your original dataset unchanged. The latter allows your analysis to be checked and replicated. Both of these are important hallmarks of doing science.

It is for these reasons that I use the R Statistical Environment when doing my statistics research. It is also for these reasons that I prefer to teach using it. I am not saying it will cure world hunger, but it will help you learn the right way to do science better than some other statistical programs—plus, it is free.

A.1 Installing R on Your Computer

As with almost any other program, installing R requires two steps: downloading it and installing it. If you wish to install R to a USB drive, that is also an option.[2]

Step 1: Download

The R program can be downloaded via the Internet. Using your web browser, go to https://cran.r-project.org/. Once there, click on the type of computer operating system (OS) you have: Linux, MacOS, or Windows.[3] On

---

[2] One would want to do this if one uses several computers and cannot be certain that R will be installed. The steps are the same as for a normal installation, with one change: Instead of selecting C://R as the destination folder, you will select the R folder on your USB drive.

[3] As of this writing, there is no option for installing R on a Chromebook or similar system. You will need to access a server that has R available. Most likely, your college/university will have an R Studio server available.

the next page, you will click on the appropriate link (depending on your OS) and download it to your desktop or some place just as convenient. The specifics are up to your browser and your computer operating system.

Step 2: Install

Once the file is on your computer, run the file (an installer) and answer the questions it asks. Usually the default selections are appropriate. The only thing you may wish to change is the destination folder. If you want to save R to your USB drive so you can bring R with you, you will have to select *that* as your destination folder.

After the installer finishes, R is installed on your computer.[4]

A.2 A Quick, Sample Session

As an example, let us do a quick, sample session that checks to make sure R is properly installed on your computer and which does no serious statistical analysis. In this session, you will start R, open a new script window, set the working directory, type in an R script, execute the script, then save the script to your current working directory.

Before you start this session, you should create a directory for your project, a folder to hold everything to do with your analysis. In reality, you *should* have a different directory for each project. This is appropriate, as it keeps your projects separate.

Step 1: Start R

If R is installed on your machine, find the icon and double-click on it. On the other hand, if it is only on your USB drive, you will need to locate the program. Likely, it is located at `USB:\\R-#.##.#\bin\` (where the `#`s are digits naming

---

[4] Actually, if you install it to your USB drive, it is not installed on the computer, *per se*. You will have to double-click the `R/bin/Rgui.exe` file each time you wish to run R from the USB drive. Doing it this way will mean you should become best friends with the `setwd()` function.

A Quick, Sample Session

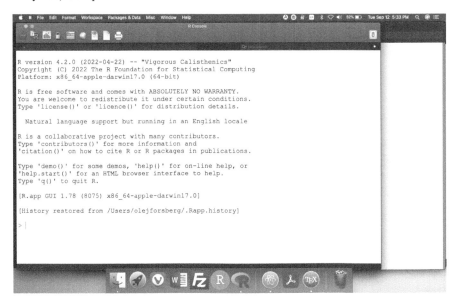

FIGURE A.1
The opening screen for R in Mac. This is for R version 4.2.0, which was released on April 22, 2022 (see the top line of the Console window). Your version may differ.

your version of R, and USB is the drive letter of your USB drive). Double-click Rgui.exe.[5] Your screen should look something like Figure A.1 (for Mac).[6]

The R window has one sub-window right now—the "R Console" window. The Console window is where all the analysis really gets done. All commands you type must eventually find their way to the Console window before R will actually execute them. However, the Console window should *not* be where you do your analysis. It is bad science to do your analysis in the Console window, since the commands you type there are lost once you press ENTER. This means *replication* of your findings will be extremely difficult and time consuming.

It is proper to type your analysis in a separate script window and send them to the Console window to be executed. Actually, "proper" is not entirely correct here, "*the only acceptable manner*" is much more accurate in

---

[5]If you installed a different version of R, then the name of the top folder will reflect that version. Also note that the specifics are for Windows installations. MacOS will use similar—though not exact—processes.

[6]Note that the primary difference between the Windows version and the MacOS version is the fact that Windows explicitly shows the Console window as a sub-window (grey background), whereas Mac does not.

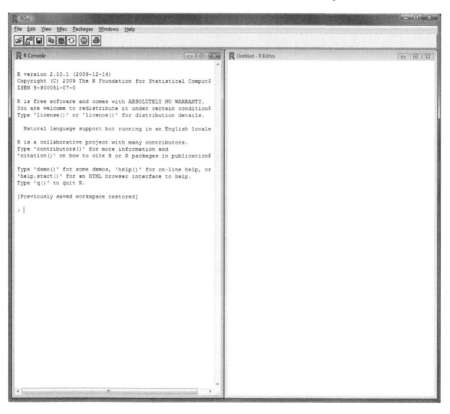

FIGURE A.2
The R window after tiling the two sub-windows (Console, left, and Script, right). *This* screen capture was done in Windows.

my opinion. Think of your script as an essay of your analysis. That may help you order it in a meaningful manner.

While there are several third-party text editors which allow you to type your analysis and send it to the Console window, R provides a text editor that is more than sufficient. The primary advantage to the built-in text editor is that it is easy to send lines of code in the Script window to the Console window to be executed—just press Command+Return when your cursor is on the appropriate line. Or, highlight the part you want to execute and press Command+Return.[7]

---

[7]This key combination is the method to use in MacOS. In Windows, you will use Ctrl+r. If you are using a third-party text editor, follow the directions provided by that vendor (which may be just copy-paste).

Step 2: Start a New Script

If you are using the R text editor on a Mac, open a new script: "File | New". You now have a second sub-window in the R window. This new window is titled "Untitled".[8] If you are working on a desktop computer, I find it helpful to tile the windows so that I can see the Console and script windows a the same time: "Windows | Tile Vertically". At this point, if you are using Windows, your R window should look similar to Figure A.2.

Step 3: Type in the Script

You have been taught to show your work from the first math class you took back in kindergarten. Continuing a long and storied tradition, you should type your script into the Script window, run it from there, and *save it* for future reference and evidence.

Type the simple script below into the Script window. The script does the following three things: creates a univariate dataset, analyzes the dataset, and plots the data. The first goal is accomplished with a single line. The second by as many lines as depth of analysis you wish to perform. The last by two lines—one for each of the two manners of graphing univariate data.

The code is as follows (make sure you type it in correctly):

```
# Basic statistics
#   script1.R

getwd()
setwd("C:/EwS2/Appendix/")

set.seed(370)
x <- runif(50000, min=10, max=20)

head(x)
tail(x)

length(x)

mean(x)
median(x)

var(x)
sd(x)
IQR(x)

min(x)
max(x)
quantile(x)
```

---

[8] On a Windows computer, the window is titled "Untitled - R Editor".

```
mean(x, trim=0.05)
quantile(x, 1:100/100)

boxplot(x)

png("histogram.png", height=4, width=4, units="in", res=600)
par(cex=0.8, cex.lab=0.8, cex.axis=0.8)
hist(x)
dev.off()
```

What does this script do? The first two lines are comments. The comment character is the octothorpe symbol, #. Anything on the line following the # is ignored by R. Commenting your script is a very good idea; it makes the script more understandable to everyone. It also allows *you* to return to the script much later, understanding your logic. It may help to think of the comments as the outline for your analysis.

The third line determines your current working directory.[9] If you save this script by pressing Ctrl+s, it will automatically be saved in that directory. If you specify a datafile or save a picture, it will be loaded or saved in that directory (unless you specify a path to the new directory). It is good practice to set, as your working directory, your project directory. That will help ensure all scripts, data files, and pictures are easily available. The fifth line sets the working directory. The one parameter is the path to the directory. That path can be relative (to the current working directory) or absolute. Line 4 provides an absolute path to the directory.[10]

It is good practice, if you are working on more than one computer, or are collaborating with other researchers, to have setwd() lines for each computer, then comment out the lines you do not need with # symbols.

The fifth line sets the random number seed, which guarantees your dataset will be the same as mine. No computer is able to produce random numbers; they produce "pseudo-random" numbers. The algorithm to produce the numbers varies from method to method, but they all are based on a number called the *seed*. Changing the value of the seed will change the series of random numbers produced. If a seed is not specifically set, then the computer will usually use the current time as the seed. This is useful if you want your trials randomized. It is not if you need to replicate the results. Thus, setting a seed allows the results in this book to correspond to the results you get, allowing you to check your work.

---

[9]In general, if a Windows command uses Ctrl, the corresponding MacOS command will use Command. There are a few exceptions to this, however.

[10]We know it is an absolute path since it starts with the drive letter.

A Quick, Sample Session

The next line creates a variable named "x" and puts 50,000 uniform random numbers, ranging from 10 to 20, in the variable x.[11] In the language of probability, x is a vector of 50000 realizations of a random variable X, such that
$$X \sim \text{UNIF}(10, 20)$$

The "r" in runif indicates its random aspect. The "unif," the continuous uniform distribution. A few other common distributions include norm (Normal or Gaussian distribution), beta (Beta distribution), and binom (Binomial distribution). Each of the other random number generator distributions have different options, see Appendix B for the useful distributions for this area of study, and well as the R Help for specifics.[12]

The next line displays the first six values of x—called the head. The tail is the *last* six values. Had I wanted to display the first 15 values, I would run head(x, 15). Doing things like helps check that your script is giving you what you expect. I expect this vector to contain 50,000 values between 10 and 20. Scanning the vector tells me that I appear to be getting something like that. The ability to examine the data is one reason why I saved it to a variable. Another reason is that I can now perform an analysis on the same data multiple times.

Line 9 displays how many elements are in x—a.k.a. the "length" or the sample size. In statistical notation, the sample size is usually represented by the variable n. As such, you may frequently see in programs a line such as:

```
n <- length(x)
```

The next several lines find the following information about the data stored in x: mean $\bar{x}$, median $\tilde{x}$, variance s^2, standard deviation s, interquartile range (IQR), minimum value, maximum value, the five-number summary (quartiles 0 through 4), the 5%-trimmed mean, and all 100 percentiles.[13]

Note the following subtle points. First, I named the data vector x, which is lower case. The vector contains 50,000 realizations of a random variable;

---

[11] The assignment operator in R is not the equals sign, =, it is the left-arrow <-, a less-than sign and a hyphen. With that said, I have yet to run into problems using an equals sign in R. This book will not hold to a standard. Sometimes, I will use <-, other times I will use =. This is to emphasize that everything should work.

[12] This is as good a place as any to introduce the R Help and R Search functions. If you know the actual command or statement, but you forget its specifics, type a question mark followed by the command in quotation marks(for example, ?"rnorm"). However, if you do not know the actual command, but you know a word close to it or what it does, type two questions marks followed by the word or words in quotation marks (for example, ??"random number").

[13] Where quartiles divide the dataset into 4 equal quarters (hence "quartile"), percentiles divide the dataset into 100 equal parts. "Equal" in this sense is number of elements. Note, however, that equal is not truly equal unless the number of elements in a dataset has certain properties; it means "approximately equals."

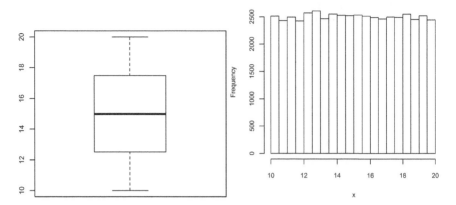

FIGURE A.3
The results of initial graphical analysis of the random uniform dataset. Left Panel: Box-and-whiskers plot. Right Panel: Histogram.

i.e., the vector is data. Second, the statistics calculated on x are lower case, too. They are realizations of random variables, themselves.

The command `boxplot(x)` plots the data as a box-and-whiskers plot (Figure A.3, left), which displays the median (heavy bar in center) and Quartiles 1 and 3 (the ends of the central box). It also displays bars at either the minimum and maximum values or, if there are outliers (signified by dots on the box-and-whiskers plot), at the data value just inside the fences.[14] There are no outliers in this dataset, so the upper bar is the maximum value in the dataset and the lower bar is the minimum value in the dataset. Your box-and-whiskers plot should look exactly like that in Figure A.3, Left Panel.

The histogram (line 23) separates the data into bins of equal width (as a default) and plots the frequency of data in each bin. Your histogram should look exactly like Figure A.3, Right Panel. As the data comes from a Uniform distribution, we expect the histogram to be flat. All bumpiness is due to the inherent randomness of sampling and the small sample size.[15]

Lines 21–24 save that histogram to your working directory as the image file **histogram.png**. Line 21 opens the file and specifies it will be 4 inches by

---

[14] The inner fences are the boundary between "acceptable" values and "outliers" in the data. The lower inner fence is $Q_1 - 1.5 \times IQR$; the upper, $Q_3 + 1.5 \times IQR$.

[15] Technically, the histogram is an approximation of the probability density function (pdf). The pdf is a "formula" for the distribution (it gives the likelihood of a specified value). Different distributions have different pdfs and (therefore) different expected histograms. As it is an approximation, the histogram will not *be* the pdf; it will only *approximate* it. Better approximation (higher precision) occurs with larger sample sizes.

A Second Example 291

4 inches with a 600 dpi resolution. Line 22 sets parameters to help ensure the histogram looks better by making the axis values smaller by 20%, the axis values by 20%, and all points by 80%. Line 23 draws the histogram to the file using those parameters, and 24 closes the image file.

Step 4: Save the script

If this were an analysis you used for your research, you would definitely want to save it. Saving the script is rather straight-forward: File | Save. If you do not see the Save option, then the active sub-window is not the script. Click on the script window and retry the save procedure.[16]

A.3 A Second Example

Let us have another example. In this example, let us analyze data imported from the Internet. In the previous example, we created our vector of data using the runif() function. Here, we will use the read.csv() function to import the data.[17]

For this example, let us demonstrate more capabilities by using a remote datafile. Its address is

   http://rfs.kvasaheim.com/data/positioningtubes.csv

The data consist of a series of length measurements of positioning tubes manufactured by a leading company. Looking at the data, you notice that the first row is the word "length." This is the name of the variable.

Here is the script.

```
###   Filename:  s2intro.R
###   Purpose:   simple analysis of the lengths
###      of positioning tubes produced by Company XY
```

---

[16] In Windows, you can also use Ctrl+s; in MacOS, Command+s.

[17] R is able to load many different data types. One of the most common data formats is the comma-separated variable format (.csv). Another is the tab-delimited format (.txt). Note that data stored in either of these two open text formats are much more accessible than data stored in proprietary formats, since any statistical program (infact, any word processing program) can open them.

With this being said, sometimes the data are stored in other formats. Thankfully, there are packages that allow you to import data in most formats. The foreign package allows one to import data in Stata's dat format, Excel's .xls or xlsx formats, and many others. Also, the readxl package is designed to make working with data in Excel formats much easier.

```
### Preamble

setwd("F:/RFS/Chapter1/")
#setwd("E:/CompanyXY/")
#setwd("C:/Consulting/CompanyXY/")

# Load a package
library(MASS)

# Access an external function
source("http://rfs.kvasaheim.com/Rfctns/means.R")

# Load and attach data
tube <- read.csv("http://rfs.kvasaheim.com/data/
    positioningtubes.csv")
attach(tube)

# What is the variable's name?
names(tube)


### Begin analysis
# Central Tendency
mean(length)
means(length,type="geometric")
means(length,type="harmonic")
median(length)

# Spread
sd(length)
var(length)
IQR(length)


### Create the graphics
boxplot(length, main="Boxplot of Positioning Tube Length",
    ylab="Positioning Tube Length [cm]")

hist(length,main="Histogram of Positioning Tube Length [cm]",
    xlab="Positioning Tube Length")
```

Note that the first few lines are comments giving information about the file, the analysis, and the data. After that, we get and set the working directory. Since I use this file on three computers, I have three different `setwd()` commands, one for each computer, with two commented out.

I need some functions from the MASS package, so I am loading it next. The MASS is one of the "about eight" packages that come with a typical installation.

Now, I know I will need to calculate the geometric mean of the data. R does not have such a function in the base package, so I will import the function from the Internet. The command is `source()`, with the filename being the argument.

A Second Example

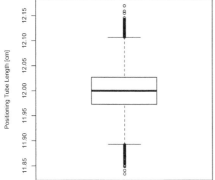
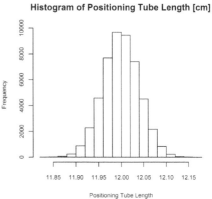

FIGURE A.4
The results of initial graphical analysis of the `positioningtube` dataset. Left Panel: Box-and-whiskers plot. Right Panel: Histogram. Note the difference in histogram shapes between this dataset and the previous. Also note how that difference is reflected in the box-and-whiskers plot.

Next, we load the data using the `read.csv()` function. Alternatively, we could have used our web browser to download the datafile to our working directory and imported it from there. With this, the entire dataset is stored in our variable `length`.

How did I know the variable was `length`? The line `names(tube)` lists the variable names in the dataframe. Be aware: The dataframe is a variable that contains variables. You can think of `tube` as a container for all variables in the dataset.

The analysis proceeds, starting at line 23. We calculate measures of central tendency and measures of spread. Finally, we produce a box-and-whiskers plot and a histogram (Figure A.4). Note how these look different than the box-and-whiskers plot and histogram from the last example. Also note that these two were displayed to your computer screen and not saved to yout working directory. If you ran the entire script in one fell swoop, you may not have seen the box-and-whiskers plot before the histogram replaced it.

Note: *In the previous example, I showed the default box-and-whiskers plot and histogram. In this example, I utilize some of the available options to make the graphics more expressive. Check to see what each option does; I leave it as an exercise for you to make the graphics look the way you prefer.*

A.4 Conclusion

In this appendix, you have learned several advantages to using the R Statistical Environment. You also learned how to download and install R on your computer (or USB drive); how to type in simple scripts; and how to use help and search functions to locate information about statements, commands, or functions in R.

The purpose of this appendix is to give you a start into being able to use R for your own statistical analyses. Since it is used exclusively in this book, I encourage you to follow along and run the code provided to get more familiar with statistics and with poll analysis.

B

Four Probability Distributions

In any statistical analysis, one needs to be aware that they are working with a sample from a larger population. As such, the sample—and any calculations based on it—are random variables with associated distributions. Because of this, it behooves us to study distributions important to the field.

This appendix does this. It examines four of the most important distributions needed for this book: Binomial, Multinomial, Beta, and Dirichlet. For those who have taken a statistics course in the past, the Binomial distribution should be vaguely familiar. The other three, not so much.

B.1 A Brief Introduction to Probability

When conducting a political poll, a sample is drawn from the sampled population. This sample is random, because two researchers performing the same sampling method will obtain two different samples. The fact that the sample is random is important because it means that any measurement on the sample will *also* be random—it is a random variable. For example, the number of successes (people voting for Candidate A) will vary from sample to sample; it is a random variable.

Random variables follow rules that specify the **likelihood** of each possible outcome. For instance, for a fair coin flipped once, the following rule describes the likelihood of each of the two possible outcomes:

$$\begin{cases} \mathbb{P}\big[\,\text{Head}\,\big] & = 0.500 \\ \mathbb{P}\big[\,\text{Tail}\,\big] & = 0.500 \\ \mathbb{P}\big[\,\text{Other}\,\big] & = 0.000 \end{cases}$$

In mathematics, this "rule" is called a function. In probability, let us call it a **probability function**. In other words, probability distributions are just functions of possible outcomes that calculate probabilities.

Note that a probability distribution can be represented in many ways. Some are more useful than others. For instance, the above probability distribution could also be represented as

$$f(x) = (0.500)^x (0.500)^{1-x}$$

where x is the number of heads flipped in one flip of the coin.

Is this formula more useful than the previous one? Maybe, maybe not. It contains exactly the same information, just presented in a different manner. What makes it useful or not is how you use it.

B.1.1 Population Parameters

Abstractly, a population parameter is a measurement on the population. This is analogous to a sample statistic, which is a measurement on the sample.

Concretely, the population parameter is what we are interested in knowing. We use our sample statistics to estimate it. In the realm of elections, the population parameter we most care about is the population proportion— π. This is the **proportion** of *all voters* who will have voted for Candidate A.

Arguably, all of inferential statistics is focused on estimating the population parameter. To help with this estimation, it helps to understand the "data-generating process"—how the data are generated and what distribution they follow. This leads to a need to understand some probability distributions. For much of our work, the Binomial distribution is all that is needed. In fact, most estimates of π still rely on just the Binomial.

However, three other distributions are useful in better understanding and estimating π. The Beta distribution is used to directly model π. The Multinomial distribution is used to estimate π_i when there are more than two candidates. The Dirichlet distribution is used to directly model all π_i when there are more than two candidates.

The Beta and Dirichlet distributions are especially important when using Bayesian methods for estimating π_i. The Beta distribution is a special case of the Dirichlet, just as the Binomial is a special case of the Multinomial.

The following explores these distributions and provides examples of their uses.

B.2 The Binomial Distribution

The Binomial distribution is the cornerstone of political polling. It is used to model the support for a single candidate in the population. The **population support** is represented by π. The random variable Y is the number of people in the sample of size n who support that candidate.

Note that Y, the number supporting Candidate A in the poll (sample), is a random variable. This means that asking a different sample will (most likely) result in a different number of supporters for Candidate A, even if π remains the same between the two polls.

For instance (trust me on the math for now), if $\pi = 0.40$ and I take two separate polls of size $n = 1000$, then the number supporting Candidate A could reasonably differ by as much as 60 people between the polls... which is equivalent to a difference of 6%... *even when the population support has not changed at all!*

As discussed in the chapter on Simple Random Sampling (Chapter 2), this possibility of such a large swing is due to the low-level of information contained in a Binomial variable. And yet, there is nothing that can be done to reduce this variability without using more information (e.g., stratified sampling, Chapter 4). It is a very part of the fabric of the universe.

B.2.1 Characteristics

Even so, let us look at the Binomial distribution—its characteristics, results, and shapes—more carefully.

- Symbol:

$$Y \sim \mathrm{BINOM}(n, \pi)$$

- R stem:

$$\texttt{binom}$$

- Likelihood function:

$$\mathcal{L}(y;\, n, \pi) = \binom{n}{y} \pi^y (1-\pi)^{n-y}$$

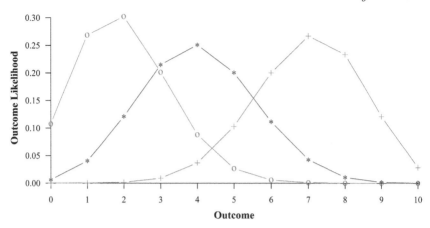

FIGURE B.1
The probability functions for a Binomial distribution using different values of π. The sample size, n, is 10 in all three cases.

Here, $B(\alpha, beta)$ is the "beta" function evaluated at those two values. It is a normalizing constant that ensures the likelihood function integrates to 1, as required.

Figure B.1 shows the probabilities of all possible outcomes when $n = 10$ and the success probability is 20% (o), 40% (*), and 70% (+). As one would expect, as the success probability increases, the most likely outcome also increases.

B.2.2 Parameters

$n \in \mathbb{N}$	number of Bernoulli trials
$\pi \in (0, 1)$	success probability in each trial
Mean:	$n\pi$
Variance:	$n\pi(1 - \pi)$

B.2.3 History

This distribution is called the Bernoulli distribution in the francophone world and the Binomial distribution elsewhere. It is called the Bernoulli distribution because Jakob Bernoulli explored it in the foundational work *Ars Conjectandi*. It is called the Binomial distribution because it makes use of the Binomial coefficient, $\binom{n}{x}$.

B.2.4 Related Distribution

One is able to calculate confidence intervals exactly using the Binomial distribution. However, because we still have vivid memories of the days in which we had to do the calculations by hand, most analysis uses the Normal approximation to the Binomial. When the sample size is "large enough," the approximation is "good." However, because the computer can perform calculations for us, it is an unnecessary approximation.

And yet, I feel compelled to give a little insight into it. So, let us be given the following:

$$X \sim \text{BINOM}(n, \pi)$$

If the number of trials, n, is "large enough," then X is *approximately* Normal:

$$X \overset{.}{\sim} \text{NORM}(n\pi, n\pi(1-\pi))$$

This is a direct result of the Central Limit Theorem.

How large is "large enough"? Usual introductory statistics text books usually suggest n is large enough if both $n\pi > 10$ and $n(1-\pi) > 10$. However, experimentation suggests that this suggestion allows an absolute error of up to 10 percentage points under the worst circumstances. Personally, if the sample size is not at least 200, then the probable errors may will exceed the difference between the candidates in a close race, making the analysis worthless.

Figure B.2 compares the BINOM(100, 0.20) distribution with the closest Normal: NORM($\mu = 20$, $\sigma = 4$). Note that the Normal does a good job of approximating the Binomial distribution. There are differences (the o points are not all on the Normal curve), but it is a good approximation.

B.2.5 Discussion

The Binomial distribution is defined as the sum of n independent and identically distributed Bernoulli distributions, each with a success probability of π. This fact makes this distribution more useful than first appears. There are five requirements for an experiment to be a Binomial experiment. Any deviation from these requirements results in a different distribution:

1. The number of trials, n, is known
2. Each trial has two possible outcomes
3. The success probability, π, is constant
4. The n trials are independent
5. The random variable is the number of successes in those n trials

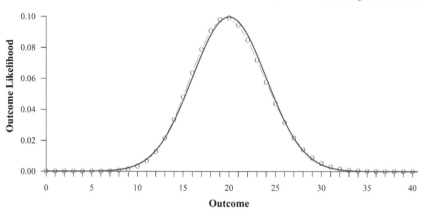

FIGURE B.2
A comparison of the Binomial distribution with the Normal. Note that the two are very close. Here, $n = 100$, $\pi = 0.20$, $\mu = 20$, and $\sigma = 4$.

In polling, n is the number of people asked; it is specified at the outset of the data collection. The "success probability" is the proportion of people in the population who are supporting Candidate A, π.

When there are multiple candidates in the election, the second requirement is usually interpreted serially. That is, the analysis will take place on each candidate individually (success = vote for Candidate A; failure = not vote for Candidate A; for each candidate).

The third requirement is that π does not change during the poll. As long as the poll is done over a short period of time and there are no significant scandals, this requirement is reasonable.

The fourth requirement is that the response of person i does not influence the response of person j.

B.2.6 An Example

For an example of the Binomial distribution, let us look at one way to determine the quality of a polling company's sampling method. The procedure is to sample from the known distribution and determine if the observed value is within the *expected bounds*.

Problem
A constitutional amendment just passed with $\pi = 0.65$ voter support. TKS Polling wishes to determine if their polling methods are properly tuned to this

The Binomial Distribution

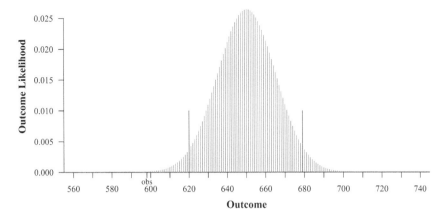

FIGURE B.3
The probability (height) of possible observations (x-axis). Note that the observed value ($x = 598$) has a very low probability of happening. Hence, we have some evidence that the sampling method employed by TKS Polling is not appropriate for this election. The endpoints of a 95% interval is provided (black segments) to illustrate how far out expected variation the observed value was.

community. To do this, they sample n voters, calculate the 95% interval, and interpret the results.

They use their sampling method to sample $n = 1000$ voters.

- In this sample, $y = 598$ reported having voted in favor of the referendum.

Solution
Let us define Y as the number of people in a sample who voted in favor of the referendum. From the information provided in this problem, the data-generating process can be represented as

$$Y \sim \text{BINOM}(n = 1000, \pi = 0.65)$$

The distribution of Y is given in Figure B.3. Note that the most likely value is 650, which is $n\pi$. Also note that we are given that the observed value is $y = 598$.

The calculated *reasonable* bounds are 620 and 679. Since the observed number of successes is outside the reasonable bounds, TKS Polling concludes that their polling methods need to be adjusted. They need to perform further exploration to determine what is wrong and if it is dependent on the type of election (referendum vs. candidate, for example).

The R command to obtain these two bounds is

```
qbinom(c(0.025,0.975), size=1000, prob=0.65)
```

Essentially, this code tells us that 95% of the time, TKS Polling should observe between 620 and 679 voters (inclusive) in favor of the referendum—if their sampling method is appropriate. That they observed a value outside that interval suggests that there is something wrong.

Code Interpretation:
The q in qbinom indicates you wish to calculate the **q**uantiles of the random variable. The binom indicates you wish to calculate those quantiles using the Binomial distribution. This function requires three pieces of information, q, n, and π. R calls the latter two size and prob.

B.2.7 A Second Example

For a more *typical* example of using the Binomial, let us estimate π and its confidence interval for a single candidate in the election.

Problem
Let us suppose TKS Polling wishes to estimate the support for Candidate A in the upcoming election. To obtain this estimate, they contact $n = 400$ people and ask if they are supporting Candidate A. In that sample, $y = 212$ claimed that they are supporting the candidate (and 188 are not).

- Calculate and interpret the 95% confidence interval for Candidate A support.

Solution
Let us define Y as the number of people in a sample who support Candidate A. From the information provided in this problem, the data-generating process can be represented as

$$Y \sim \text{BINOM}(n = 400, \ \pi)$$

with π unknown.

The observed value is $y = 212$. Using the computer, we have a 95% confidence interval from 47.9% to 58.0%. Since this interval includes values less than 50%, we cannot conclude (at this level) that the candidate will win (get at least 50% of the vote).

The Binomial Distribution 303

Technically, we would say that *under repeated experiments*, 95% of such confidence intervals thus obtained would contain the true level of support for Candidate A, π. We do not know if this particular confidence interval contains it. We just know that 95% of all confidence intervals do.

This is why we use the term "confidence" instead of "probability." A confidence interval is not a probability statement about π. A confidence interval provides *reasonable* values of π, given the data.

By the way, the code to obtain the endpoints of the confidence interval is

```
binom.test(x=212, n=400)
```

Code Interpretation:
The `binom.test` function can be used to estimate the population proportion given the number of observed successes, `x`, and the number of trials, `n`. The function can also be used to test a hypothesis about π by providing its hypothesized value. If no such value is provided, then R assumes it is testing the claim $\pi = 0.500$.

B.2.8 Comparing Two Candidates

Much political polling is done in elections with two major candidates or positions. In such cases, there is an unfortunate tendency to directly compare the two candidates to each other. **Do not do this.** If there are only two positions, A and B, then we know the support for B if we know the support for A. Thus, there is really only one Binomial distribution involved.

To illustrate this, let us return to the previous example. We estimated the support for Candidate A to be between 47.9% and 58.0%. We could just as easily estimated the support for Candidate B using the same data:

```
binom.test(x=188, n=400)
```

which would give a 95% confidence interval from 42.0% to 52.1%. Note that this is just 100% minus the endpoints of the confidence interval for Candidate A. That is, support for Candidate A is *not* independent of support for Candidate B. They are, in fact, perfectly correlated. In such cases, it is correct to show a single estimate and confidence interval, as in Figure B.4. I leave it to the graphic artists amongst us to make the graphic look better.

What is this Binomial Test?

In the text, I discussed the Wald test as a method for estimating π from the data. That test requires a rather large sample because it uses the Normal distribution to approximate the Binomial (Section B.2.4). The Binomial test can be used for any sample size.

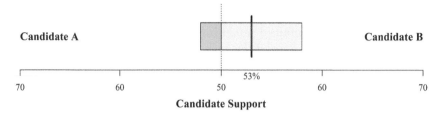

FIGURE B.4
An appropriate illustration of the estimated support level for a race consisting of just two candidates. This clearly illustrates that π could be on either side of the 50% line.

Note, however, that as the sample size increases, the difference between the two tests goes to zero. With a sample size of at least 400, there is precious little difference between the two methods. Thus, think of this as a quick introduction to the Binomial test.

B.2.9 Comparing More Than Two Candidates

This leads us to the next issue (and distribution). The Binomial distribution can be used to estimate the support level for each candidate *individually*. However, it should *not* be used to compare support for the candidates. Section 7.2 discussed this at length. The following provides another way of seeing the issue. It is a pernicious problem that pervades much statistics. So, spending a second section covering it is a good use for the ink (or toner).

As an example, let us have an election among three candidates. To estimate support for each, TKS Polling contacts $n = 400$ people and asks each their preference. In the sample, Candidates A, B, and C receive 160, 140, 100 votes, respectively.

TKS Polling "rightly" produces the graphic in Figure B.5. In that graphic, we see that the confidence intervals for Candidate A and B overlap. This rightly suggests that the poll cannot claim one is ahead of the other. The confidence interval for Candidate C does not overlap with either of the other two candidates. This suggests that Candidate C is not currently ahead—wrongly.

Note that this graphic shows three completely *different* Binomial distributions (technically, the confidence intervals resulting from them). We are 95% confident that Candidate A support is between 35% and 45%; Candidate B, between 30% and 40%; and Candidate C, between 21% and 29%. These are correct statements about each candidate *individually*.

The Binomial Distribution 305

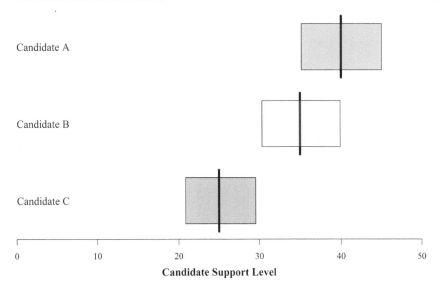

FIGURE B.5
The observed proportions (vertical segments) and the 95% confidence intervals for each of three candidates in the race. Note that this improperly invites comparison between candidates.

However, given Figure B.5, one naturally wants to compare candidate support across candidates. This would require another distribution... one that allows for three possible outcomes for each person asked. This is the Multinomial distribution. However, to make the mathematics easier, one can adjust the confidence level to take into consideration the number of comparisons made, c. The typical adjustment is the Bonferroni adjustment. Instead of using a 95% confidence interval, one uses a $1 - 0.05/c$ level. This is a conservative adjustment (see Section 7.2.1).

Thus, for the example, we would create 98.33% confidence intervals and interpret them as if they were 95% confidence intervals. This is statistically tractable and well understood in the literature. It is also in line with the usual method for presenting such a race (Figure B.5). Unfortunately, this correction produces confidence intervals that are wider than they should be (see Figure B.6), thus this graphic cannot be used to estimate a single candidate's support level.

The "correct" way to deal with races with multiple candidates is to use the Multinomial distribution in lieu of the Binomial distribution. The drawback to this correct solution is that representing the results is quite difficult and require the reader to re-think how the information is presented. For instance, if there are three candidates, then one can use an equilateral triangle to represent the polling results (Figure 7.3 on page 157 and B.7 below). However, if there

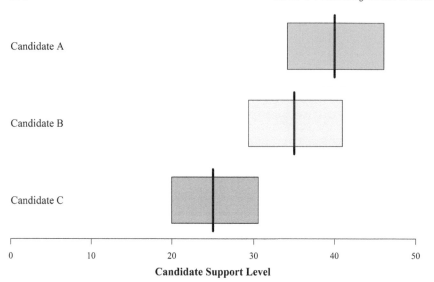

FIGURE B.6
The observed proportions (vertical segments) and the 95% confidence intervals for each of three candidates in the race. Note that this more correctly invites comparison between candidates because of the Bonferroni adjustment.

are four candidates, an equilateral tetrahedron is needed. Representing such a three-dimensional figure in two dimensions is a challenge, to say the least.

More than you want to know

There are two things going on here. The first is that confidence intervals (regions) based on the Normal distribution are always ellipsoids. The second is that these ellipsoids are projected onto a lower dimension based on the number of free parameters.

For instance, in the case of two candidates, the confidence region is an ellipse. It is projected down into one dimension because the two support levels must sum to 1: $\pi_a + \pi_b = 1$; there is only one free parameter because $\pi_b = 1 - \pi_a$. As such, when there are two candidates, it is acceptable to represent the poll results as in Figure B.4.

When there are three candidates, the ellipsoid is three-dimensional, but projected down in to two dimensions as an ellipse because the support for the three candidates must sum to 1: $\pi_a + \pi_b + \pi_c = 1$; there are only two free parameters because $\pi_c = 1 - \pi_a - \pi_b$. Thus, when there are three candidates, it is more appropriate (from a mathematical perspective) to illustrate the confidence region in two dimensions, as in Figure B.7.

The Binomial Distribution 307

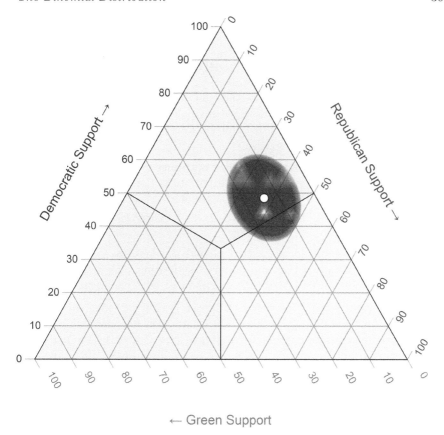

← Green Support

FIGURE B.7
A ternary graphic of support for a three-candidate race, with a 95% confidence region shown. This graphic is for first-past-the-post elections (plurality).

Figure B.7 illustrates a 95% confidence region for a three-person race in which TKS Polling surveyed $n = 600$ people, 291 of whom stated they supported the Democratic candidate, 224 the Republican, and 85 the Independent. Note that the observed proportions (the white dot) is in the Democratic region (top), indicating that the Democratic candidate received a plurality of the support *in the poll*. The confidence region includes points in both the Democratic and Republican regions. This indicates that this poll suggests that the race is a toss-up between the Democrat and Republican. That the confidence region is not in the Independent region at all indicates that the Independent candidate will not win—according to the poll.

B.3 The Multinomial Distribution

In the previous section, we looked at the Binomial distribution, which should only be used when there are two positions or candidates in the election. If there are more than two, then another distribution must be used. That distribution is the Multinomial distribution.

When there are more than two possible outcomes in a single population, it is more correct to use the Multinomial distribution. For instance, an election with three candidates. Or a poll in which there are two candidates... and an option for undecided. In both of these cases, each person polled can answer in one of three ways.

Why don't we use this for all statistical analysis? It seems that there is one major difficulty with using the Multinomial distribution. The first is that it is more difficult to illustrate, either with a graphic or with confidence intervals.

In the previous section (as well as Chapter 7), we were able to illustrate a race with three candidates. To illustrate a race with *four* candidates, a triangular pyramid must be drawn—three dimensions. Currently, the paradigm for presenting results is the piece of paper—two dimensions. With advances in web technology, there may come a time in which we are able to represent a three dimensional graphic in its entirety. But, that would only work for a four-person (or party) race. If there are more than four parties, then the number of dimensions would increase, too.

B.3.1 Characteristics

Even so, let us look at the Binomial distribution—its characteristics, results, and shapes—more carefully.

- Symbol:
$$Y \sim \text{MULTINOM}(n, \pi_1, \pi_2, \ldots, \pi_k)$$

- R stem:
```
multinom
```

- Likelihood function:
$$\mathcal{L}(y;\ n, \pi_1, \pi_2, \ldots, \pi_k) = \frac{n!}{y_1!\ y_2!\ \cdots\ y_k!}\ \pi_1^{y_1}\ \pi_2^{y_2}\ \cdots\ \pi_k^{y_k}$$

The Beta Distribution 309

Note that $y_1 + y_2 + \cdots + y_k = n$ and $\pi_1 + \pi_2 + \cdots + \pi_k = 1$. Thus, we see that the Binomial is a *special case* of the Multinomial, one where there are only two possible outcomes, 1 and 2, with the number of successes for each of these two outcomes being y_1 and $y_2 = n - y_1$ and the success probabilities being π_1 and $\pi_2 = 1 - \pi_1$.

B.3.2 Parameters

$n \in \mathbb{N}$	number of trials
$\pi_i \in (0, 1)$	probability of obtaining outcome i in each trial
Mean for outcome i:	$n\pi_i$
Variance for outcome i:	$n\pi_i(1 - \pi_i)$

Thus, we see that a lot of what we learned in the Binomial section applies to the Multinomial—suitable generalized to multiple outcomes.

B.4 The Beta Distribution

The previous two distribution modeled the number of successes in the sample. However, we were able to use them to estimate the support in the population. The Beta distribution can directly model the support in the population. In doing this, we leave the realm of frequentist statistics and become Bayesians. Check out Chapter 5 for more on the difference between the two types of statistics.

B.4.1 Characteristics

Even so, let us look at the Binomial distribution—its characteristics, results, and shapes—more carefully.

- Symbol:

$$Y \sim \text{BETA}(\alpha, \ \beta)$$

- R stem:

```
beta
```

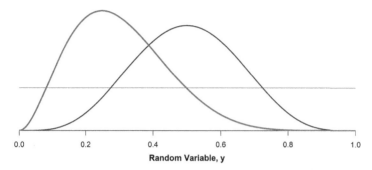

FIGURE B.8
The probability density function for three different Beta distributions. The thin curve is for BETA(1, 1); the medium curve, BETA(5, 5); the thick curve, BETA(3, 7).

- Likelihood function:

$$\mathcal{L}(y;\ \alpha,\ \beta) = B(\alpha, \beta)\ y^{\alpha-1}\ (1-y)^{\beta-1}$$

Figure B.8 shows the likelihood for three Beta distributions. Note the flexibility of the distribution. The thin red line is equivalent to a standard Uniform distribution, one in which all probabilities are equally likely. The other two curves are for different expected values for the same sample size of $n = 10$.

B.4.2 History

The first record of the Beta distribution is in Bayes' paper.[Bay63] He discovered the Beta distribution as the distribution for success probabilities in Bernoulli trials. That is how we currently use it for the most part, to model π.[JK97] It is used in other disciplines; however, those uses still treat it as a way to model some probability or population proportion (e.g. population genetics).[BN95]

B.4.3 Parameters

$\alpha > 0$	one shape parameter ("number of successes")
$\beta > 0$	another shape parameter ("number of failures")
Mean:	$\dfrac{\alpha}{\alpha + \beta}$
Variance:	$\dfrac{\alpha}{\alpha + \beta} \dfrac{\beta}{(\alpha + \beta)(\alpha + \beta + 1)}$

The Dirichlet Distribution

If we interpret α and β as the numbers of successes and failures (respectively), then the expected value is equivalent to the number of successes divided by the sample size. Make sure this makes sense to you.

B.4.4 An Example

Since the most important use of the Beta distribution is in Bayesian analysis, let us see an example of this use. For background information on Bayesian analysis and conjugate priors, see Section 3.3 and Chapter 5.

TKS Polling conducts a poll to determine who will win the Galesburg mayoral election. In the sample of $n = 400$, a total of $x = 213$ people stated they supported Peter Schwartzman, and 197 supported the other candidate.

Using a uniform conjugate prior, BETA$(1, 1)$, and this data, the posterior distribution is BETA$(294, 108)$. According to this poll, the probability that Schwartzman wins is 79%. In fact there is a 95% probability that he will receive between 47% and 57% of the vote.

Here are the calculations:

```
1-pbeta(0.50, 213, 197)
qbeta(c(0.025,0.975), 213, 197)
```

B.4.5 Continuing the Example

Shortly after taking that poll, TKS Polling ran a second one. In that second sample of $n = 400$ people, the number stating they were going to vote for Schwartzman was $x = 243$.

If we use the previous results, then the posterior distribution will be BETA$(458, 342)$. According to this poll and the previous, the probability that Schwartzman wins is now over 99%. In fact there is a 95% probability that he will receive between 54% and 61% of the vote.

Here are the calculations:

```
1-pbeta(0.50, 458, 342)
qbeta(c(0.025,0.975), 458, 342)
```

B.5 The Dirichlet Distribution

The Dirichlet distribution is an extension of the Beta distribution to more than just two outcomes. Where the Beta distribution is used as the conjugate prior for the Binomial, the Dirichlet is used as the conjugate prior for the

Multinomial. Thus, everything you learned about the Beta can apply to the Dirichlet.

B.5.1 Characteristics

With that, let me give the specifics of the Dirichlet distribution here:

- Symbol:
$$Y \sim \text{DIR}(\alpha_1, \alpha_2, \ldots \alpha_k)$$

- R stem:
```
dirichlet
```

- Likelihood function:
$$\mathcal{L}(y; \alpha_1, \alpha_2, \ldots \alpha_k) = B(\alpha_1, \alpha_2, \ldots \alpha_k)\, y_1^{\alpha_1 - 1}\, y_2^{\alpha_2 - 1} \cdots y_k^{\alpha_k - 1}$$

B.5.2 History

Peter Gustav Lejeune Dirichlet was a German mathematician with his fingers in a lot of subdisciplines. While this distribution was not *actually* created by Dirichlet. It was, however, based on one of his Dirichlet Integrals. The main purpose of the Dirichlet distribution is to extend the Beta distribution to multiple dimensions. This is our use, too.

Beyond this, there is not much else to say about the Dirichlet distribution.

B.5.3 A First Example

The examples in the Beta distribution section work perfectly if there are two candidates in the race. Let us look at an example where there are six parties running for a single seat.

This example comes from the United Kingdom parliament, which consists of 650 members elected from single-member districts. A candidate wins the district if they receive the most votes, not necessarily a majority of them. This type of election is termed SMP: single-member plurality.

In the 2019 election, St. Ives in beautiful Cornwall had five parties contest the seat: Common People, Conservative, Green, Labour, Liberal, and Liberal Democrats. To determine the current state of the election, TKS Polling surveyed $n = 500$ people in St. Ives. Their preferences were, in order: 3, 248, 10, 18, 16, and 205.

The Dirichlet Distribution

Using a uniform (flat) prior, we know that the posterior will also be a Dirichlet distribution. In fact, it will be

$$\pi \sim \text{DIR}(4, 249, 11, 19, 17, 206)$$

Now, with the posterior distribution, we can calculate some interesting probabilities:

- The probability that the Conservatives will receive a majority of the votes cast is 36.1%.

- The probability that the Liberal Democrats will receive a majority of the votes cast is about 0.

- Most importantly, because this is a first-past-the-post election, the probability that the Conservatives will hold the seat is 97.8%.

Unfortuantely, R does not come with a Dirichlet-related function. Thus, you will need to install a package that has it. I tend to use the gtools package. Once you have that package loaded, this code will give you the above results.

```
library(gtools)

prior = c(1,1,1,1,1,1)
observed = c(3, 248, 10, 18, 16, 205)
posterior = prior+observed

y = rdirichlet(1e6, posterior)

colnames(y) = c("Common People","Conservative","Green","
    Labour","Liberal","Liberal Democrats")

# Calculations
mean(y[,"Conservative"]>=0.50)
mean(y[,"Liberal Democrats"]>=0.50)
mean(y[,"Conservative"]>=y[,"Liberal Democrats"])
```

I used the colnames function here just to make the analysis more transparent. Without that function, the calculations would be the less-clear

```
mean(y[,2]>=0.50)
mean(y[,6]>=0.50)
mean(y[,2]>=y[,6])
```

B.5.4 Continuing the First Example

Let us continue the above example. TKS Polling conducts a second poll in the St. Ives constituency. In this second poll, the preferences were, in order: 1, 203, 4, 27, 18, and 247.

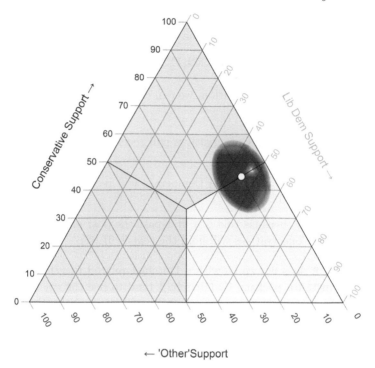

FIGURE B.9
A graphic showing the predicted support level for Conservatives, Liberal Democrats, and "other" candidates. The credible region is in both the Conservative and the Liberal Democratic regions. Thus, these polls do not indicate which party will win the election.

If we use the above posterior distribution as the *prior* for this problem, we have the following posterior distribution given the new data:

$$\pi \sim \text{DIR}(5, 452, 15, 46, 35, 453)$$

Now, with the new posterior distribution (Figure B.9), we can calculate the most important probability.

- The probability that the Conservatives will hold the seat is 48.6%.

Thus, with the new data (poll), the probability that the Conservatives will win the election in St. Ives is essentially 50-50, a coin-flip.

What a difference data makes.

B.6 End of Appendix Materials

B.6.1 R Functions

In this appendix, we were introduced to several R functions dealing with probability distributions that may be very useful in the future.

Binomial Distribution

R uses the parameterization where size is the number of trials and prob is the success probability. The variable represents the number of successes in the sample.

dbinom(x, size,prob) Returns the probability for an x-value according to the specified Binomial distribution; it calculates $\mathbb{P}[\,X = x\,]$.

pbinom(x, size,prob) Returns the cumulative probability for an x-value according to the specified Binomial distribution; it calculates $\mathbb{P}[\,X \leq x\,]$.

qbinom(p, size,prob) Returns the quantile (percentile) according to the Binomial distribution specified; it calculates x_p such that $\mathbb{P}[\,X \leq x_p] = p$.

rbinom(n, size,prob) Returns n random numbers from the specified Binomial distribution.

Multinomial Distribution

R uses the parameterization where size is the number of trials and prob is a vector of success probabilities. The elements of prob must sum to 1. The variable represents the number of successes in each category (length of prob).

dmultinom(x, size,prob) Returns the probability for an x-value according to the specified Multinomial distribution; it calculates $\mathbb{P}[\,\mathbf{X} = \mathbf{x}\,]$.

rmultinom(n, size,prob) Returns n random numbers from the specified Multinomial distribution.

Beta Distribution

R uses shape1 and shape2 for α and β.

dbeta(x, shape1,shape2) Returns the likelihood of an x-value according to the specified Beta distribution.

pbeta(x, shape1,shape2) Returns the cumulative probability for an x-value according to the specified Beta distribution; it calculates $\mathbb{P}[\,X \leq x\,]$.

qbeta(p, shape1,shape2) Returns the quantile (percentile) according to the Beta distribution specified; it calculates x_p such that $\mathbb{P}[\,X \leq x_p] = p$.

rbeta(n, shape1,shape2) Returns n random numbers from the specified Beta distribution.

Dirichlet Distribution

R uses alpha for the main parameter. This alpha is actually a vector of values, one per possible outcome. So, if you are working with a three-candidate election, you would provide three values in alpha.

ddirichlet(x, alpha) Returns the likelihood for an x-value according to the specified Dirichlet distribution. Here, alpha is a vector of values such that its sum is the sample size n.

rdirichlet(n, alpha) Returns n random numbers from the specified Dirichlet distribution.

Bibliography

[AAP24a] AAPOR. Standards and ethics. Internet, 2024. `https://aapor.org/standards-and-ethics/#aapor-code-of-professional-ethics-and-practices`.

[AAP24b] AAPOR. Transparency initiative. Internet, 2024. `https://aapor.org/standards-and-ethics/transparency-initiative/`.

[AAP24c] AAPOR. Who we are. Internet, 2024. `https://aapor.org/about-us/who-we-are/`.

[AC98] Alan Agresti and Brent A. Coull. Approximate is better than "exact" for interval estimation of binomial proportions. *The American Statistician*, 52:119–126, 1998.

[Afg09a] Afghanistan. Polling regulations. Internet, 2009. `http://iec.org.af/pdf/legalframework/regulations/eng/RegulationOnPolling.pdf`.

[Afg09b] Afghanistan. Presidential and provincial council elections. Internet, 2009. `http://www.iec.org.af/results_2009/`.

[Age10] Agence France-Presse. Constitutional body names Gbagbo I.Coast election winner. Internet, 3 December 2010. `http://www.google.com/hostednews/afp/article/ALeqM5h1lqqW8eecnVcdL82ggEDWQRliOQ?docId=CNG.a5fc0e83efff72426ce88ff122d81b07.751`.

[Age15] Agence France-Presse. Sri Lanka monitors fear voter intimidation before election. Internet, 4 January 2015. `https://www.dailymail.co.uk/wires/afp/article-2896128/Sri-Lanka-monitors-fear-voter-intimidation-election.html`.

[Aho14] Ken A. Aho. *Foundational and Applied Statistics for Biologists*. Chapman & Hall / CRC Press, 2014.

[Al 19] Al Jazeera. Sri Lanka bombings: All the latest updates. Internet, 2 May 2019. `https://www.aljazeera.com/news/2019/5/2/sri-lanka-bombings-all-the-latest-updates`.

[Alb09] Jim Albert. *Bayesian Computation with R*. Use R! Springer Press, 2nd edition, 2009.

[Ara23]   Ozdemir Arastirma. Genel milletvekili seçimleri siyasi partilerin oy tercihleri. Internet, 2023. [Online; accessed 5-March-2024].

[Bay63]   Thomas Bayes. An Essay towards Solving a Problem in the Doctrine of Chances. By the Late Rev. Mr. Bayes, F. R. S. Communicated by Mr. Price, in a Letter to John Canton, A. M. F. R. S. *Philosophical Transactions of the Royal Society of London*, 53:370–418, 1763.

[BBC10a] BBC News. Sri Lanka MPs vote in sweeping powers for president: Analysis. Internet, 8 September 2010. https://www.bbc.com/news/world-south-asia-11225723.

[BBC10b] BBC News. Sri Lanka president wins re-election— state TV. Internet, January 2010. http://news.bbc.co.uk/2/hi/south_asia/8482270.stm.

[BBC10c] BBC News. Sri Lanka presidential votes being counted. Internet, 26 January 2010. http://news.bbc.co.uk/2/hi/south_asia/8478386.stm.

[BBC11a] BBC News. Ivory Coast: Gbagbo held after assault on residence. Internet, 11 April 2011. http://www.bbc.co.uk/news/world-africa-13039825.

[BBC11b] BBC News. Sri Lanka's jailed ex-army chief Fonseka given new term. Internet, 18 November 2011. http://www.bbc.com/news/world-asia-15787672.

[Bel03]   Imogen Bell. *Central and South-Eastern Europe 2003*. Psychology Press, 2003.

[Ben38]   Frank Benford. The law of anomalous numbers. *Proceedings of the American Philosophical Society*, 78(4):551–572, March 1938.

[BN95]   David J. Balding and Richard A. Nichols. A method for quantifying differentiation between populations at multi-allelic loci and its implications for investigating identity and paternity. *Genetica*, 96:3–12, June 1995.

[Boz21]   Mahmut Bozarslan. Turkey's kurds revive fight for language rights. Internet, 2021. https://www.al-monitor.com/originals/2021/11/turkeys-kurds-revive-fight-language-rights.

[BSVM14] Nancy D. Berkman, P. Lina Santaguida, Meera Viswanathan, and Sally C Morton. The empirical evidence of bias in trials measuring treatment differences. Technical Report 14-EHC050-EF, Agency for Healthcare Research and Quality (US), September 2014.

[BWL22]   Ben Bolker, Gregory R. Warnes, and Thomas Lumley. *gtools: Various R Programming Tools*, 2022. R package version 3.9.4.

Bibliography

[Car88]   Charles A. P. N. Carslaw. Anomalies in income numbers: Evidence of goal oriented behavior. *The Accounting Review*, 63(2):321–327, April 1988.

[Cen10]   Centre for Monitoring Election Violence (CMEV). Final report on election related violence and malpractices: Presidential Election 2010. Internet, February 2010. https://cmev.files.wordpress.com/2010/07/presidential-election-2010-final-report.pdf.

[Cen20]   Central Election Commission. 2020 Presidential and Vice Presidential Election. Internet, 2020. https://web.cec.gov.tw/english/cms/pe/32471.

[Cen24]   Central Intelligence Agency. Bolivia - the world factbook. Internet, 2024. https://www.cia.gov/the-world-factbook/countries/bolivia/.

[CG07]   Wendy Tam Cho and Brian Gaines. Breaking the (Benford) law: Statistical fraud detection in campaign finance. *The American Statistician*, 61(3):218–223, August 2007.

[Che18]   Brahma Chellaney. Sri Lanka— democracy in danger. Internet, 13 November 2018. https://asia.nikkei.com/Opinion/Sri-Lanka-democracy-in-danger.

[Chr02]   Ronald Christensen. *Plane Answers to Complex Questions: The Theory of Linear Models*. Springer, 3rd edition, 2002.

[Coh16]   Nate Cohn. We Gave Four Good Pollsters the Same Raw Data. They Had Four Different Results. Internet, 20 September 2016. https://www.nytimes.com/interactive/2016/09/20/upshot/the-error-the-polling-world-rarely-talks-about.html.

[Coh19]   Nate Cohn. No one picks up the phone, but which online polls are the answer? Internet, 2 July 2019. https://www.nytimes.com/2019/07/02/upshot/online-polls-analyzing-reliability.html.

[Col22a]   Collier County Supervisor of Elections. Collier County Supervisor of Elections. Internet, 2022.

[Col22b]   Collier County Supervisor of Elections. May district demographic analysis. Internet, 2022.

[Com10]   Commission Electorale Independante. Election Results, Côte d'Ivoire 2010 President (CEI). Internet, 2010. https://www.cei-ci.org/redirect/web/file/uploads/e6c737_resultats-du-second-tour.pdf.

[Dev08]   Tom M. Devine. *Scotland and the Union 1707–2007*. Edinburgh University Press, 2008.

[Don17] Dong-A Ilbo. Moon Jae-in 40.2%, Ahn Cheol-soo 19.9%, Hong Jun-pyo 17.7%. Internet, 3 May 2017. http://www.donga.com/ISSUE/2017president/News?gid=84188748&date=20170503.

[Dor16] USC Dornsife. The 2016 usc dornsife / la times presidential election poll. Internet, 2016. https://election.usc.edu/2016/.

[Dor20] USC Dornsife. The 2020 usc dornsife / la times presidential election poll. Internet, 2020. https://election.usc.edu/index.html.

[DS15] Robert A. Dahl and Ian Shapiro. *On democracy*. Yale University Press, New Haven, 2015.

[Duv23] Gazete Duvar. Emek ve Özgürlük İttifakı'nda tam uzlaşı sağlandı [full consensus reached in the labor and freedom alliance]. Internet, 2023. https://www.gazeteduvar.com.tr/emek-ve-ozgurluk-ittifakinda-tam-uzlasi-saglandi-haber-1609870.

[EF18] Uduak-Obong I. Ekanem and Ole J. Forsberg. An analysis of the 2015 Nigerian presidential election. *PURSUE*, 1(2):5–20, 2018. http://www.pvamu.edu/pursue/wp-content/uploads/sites/155/2018/06/vol1-issue2.pdf.

[EK22] Samson Ellis and Adrian Kennedy. Xi's suppression of hong kong democracy pushes Taiwan further from China. https://www.japantimes.co.jp/news/2022/07/04/asia-pacific/xi-suppression-hong-kong-taiwan/, 4 July 2022.

[Eld19] Mine Elder. Turkey. In Ellen Lust, editor, *The Middle East*, pages 695–728. CQ Press, 2019.

[Ele94] Election Commission of Sri Lanka. 1994 Sri Lanka Presidential Election Results. Internet, 1994. https://results.elections.gov.lk/.

[Ele99] Election Commission of Sri Lanka. 1999 Sri Lanka Presidential Election Results. Internet, 1999. https://results.elections.gov.lk/.

[Ele05] Election Commission of Sri Lanka. 2005 Sri Lanka Presidential Election Results. Internet, 2005. https://results.elections.gov.lk/.

[Ele10] Election Commission of Sri Lanka. 2010 Sri Lanka Presidential Election Results. Internet, 2010. https://results.elections.gov.lk/.

[Ele15] Election Commission of Sri Lanka. 2015 Sri Lanka Presidential Election Results. Internet, 2015. https://results.elections.gov.lk/.

Bibliography 321

[Ele18a] Election Commission of Sri Lanka. Election Commission. Internet, 31 December 2018. http://elections.gov.lk/web/en/election-commission/.

[Ele18b] Election Commission of Sri Lanka. Presidential election results. Internet, 31 December 2018. https://results.elections.gov.lk/.

[Ele19] Election Commission of Sri Lanka. 2019 Sri Lanka Presidential Election Results. Internet, 2019. https://results.elections.gov.lk/.

[Ele24] Election Commission of Bhutan. General Election Results, 2024. Internet, 2024. https://www.ecb.bt/GEResults2024/.

[Eme] David Emery. Internet. https://www.snopes.com/fact-check/stalin-vote-count-quote/.

[Erl16] Steven Erlanger. Britain votes to leave e.u.; cameron plans to step down. https://www.nytimes.com/2016/06/25/world/europe/britain-brexit-european-union-referendum.html, June 2016.

[ETt23] ETtoday. ET poll/Zhao Shaokang breaks through 30%, "Hou Kang" support surges to 32.5%. Internet, 25 November 2023. https://www.ettoday.net/news/20231125/2630294.htm.

[Eur04] European Union Election Observation Mission. Sri Lanka Parliamentary Elections 2 April 2004. Internet, 2 April 2004. http://www.eods.eu/library/FR%20SRI%20LANKA%202004_en.pdf.

[Far39] Richard Farley. *Tables of Logarithms*. Taylor and Walton, 1839.

[Fed22] Federal Election Commission. Federal elections 2020. https://www.fec.gov/resources/cms-content/documents/federalelections2020.pdf, October 2022. [Online; accessed 20-March-2024].

[For14] Ole J. Forsberg. *Electoral Forensics: Testing the "free and fair" claim*. Dissertation, Oklahoma State University, Stillwater, OK, 2014.

[For20] Ole J. Forsberg. *Understanding Elections through Statistics: Polling, Prediction, and Testing*. CRC Press, New York, 2020.

[Fre17] Freedom House. Sri Lanka. Internet, 2017. https://freedomhouse.org/country/sri-lanka/freedom-world/2017.

[GAD20] GADM. GADM Shapefiles, version 3.6. Internet, 2020. http://www.gadm.org/.

[GAD24] GADM. GADM Shapefiles, Version 4.1. Internet, 2024. http://www.gadm.org/.

[Gal09]   Peter W. Galbraith. How the Afghan Election was Rigged. *Time Magazine*, 174(15), 19 October 2009. http://www.time.com/time/magazine/article/0,9171,1929210,00.html.

[Gal17]   Gallup-Korea. Gallup Korea Daily Opinion No.257. Internet, May 2017. http://gallupkorea.blogspot.com/2017/05/2572017-5-1.html.

[GCS+13]   Andrew Gelman, John B. Carlin, Hal S. Stern, David B. Dunson, Aki Vehtari, and Donald B. Rubin. *Bayesian Data Analysis*. Chapman & Hall/CRC Press, 3rd edition, 2013.

[Gha12]   Ghana. *C.I.75 Public Elections Regulations*. Electoral Commission of Ghana, 2012. http://www.judicial.gov.gh/images/stories/File/C.I.%2075.pdf.

[Gha20]   Ghana. *Elections 2020: What You Should Know*. Electoral Commission of Ghana, 2020. https://ec.gov.gh/wp-content/uploads/2020/12/WhatsApp-Image-2020-12-06-at-11.44.39-2.jpeg.

[Glo12]   Global Investment Center. *Gambia Business Law Handbook*, volume 1: Strategic Information and Basic Laws. International Business Publications, 2012.

[Gos08]   "Student" [William Sealy Gosset]. The Probable Error of a Mean. *Biometrika*, 6(1):1–25, March 1908.

[Gro00]   Laura Gross. Democracy continues, Sri Lanka style... Internet, 24 October 2000. https://tamilnation.org/tamileelam/democracy/001024laura.htm.

[Gul13]   Gulf-Times. Lankan Tamils vote amid charges of intimidation. Internet, 21 September 2013. https://www.gulf-times.com/story/366398/Lankan-Tamils-vote-amid-charges-of-intimidation.

[Han17]   Hankook Research. [Population poll] Moon Jae-in 40.2%... Hong, An, 'Error Range'. Internet, 2 May 2017. http://news.jtbc.joins.com/article/article.aspx?news_id=NB11463059.

[Hel16]   Tobu Helm. EU referendum: youth turnout almost twice as high as first thought. Internet, 10 July 2016. https://www.theguardian.com/politics/2016/jul/09/young-people-referendum-turnout-brexit-twice-as-high.

[Hen10]   Rohini Hensman. Sri Lanka becomes a dictatorship. *Economic and Political Weekly*, 45(41):41–46, 2010.

[Hil95]   Theodore P. Hill. A statistical derivation of the significant-digit law. *Statistical Science*, 10(4):354–363, November 1995.

[Hum15]   Human Rights Watch. Sri Lanka: Violence, intimidation threaten vote, Jan 2015.

[Hyd11]   Susan D. Hyde. *The Pseudo-Democrat's Dilemma: Why Election Observation Became an International Norm.* Cornell University Press, 2011.

[IAR22]   Tara John Iqbal Athas, Rukshana Rizwie and Hannah Ritchie. Protesters storm Sri Lanka's prime minister's office, as president flees country without resigning. Internet, July 13 2022. https://edition.cnn.com/2022/07/12/asia/sri-lanka-crisis-gotabaya-rajapaksa-airport-intl/index.html.

[Igi20]   Ruth Igielnik. Men and women in the u.s. continue to differ in voter turnout rate, party identification. Internet, 2020. [Online; accessed 26-March-2024].

[Ind09]   Independent Electoral Commission. Election Results, Afghanistan 2009 President. Internet, 2009. http://www.iec.org.af/results/en/leadingCandidate.html.

[Ind10]   Independent Electoral Commission. Elections Présidentielles 2010. Internet, 3 November 2010. https://www.abidjan.net/elections/presidentielle/2010/Resultats/2emetour/.

[Ind16]   Independent Electoral Commission. Presidential Election Results 1st December 2016. Internet, 2016. https://iec.gm/download/presidential-election-results-1st-december-2016/.

[JK97]   Norman L. Johnson and Samual Kotz. *Leading Personalities in Statistical Sciences: From the Seventeenth Century to the Present.* Wiley Series in Probability and Statistics. Wiley, 1997.

[kdP17]   Fondation-Institut kurde de Paris. The kurdish population. Internet, 2017. https://www.institutkurde.org/en/info/the-kurdish-population-1232551004.

[KMK+16]   Courtney Kennedy, Andrew Mercer, Scott Keeter, Nick Hatley, Kyley McGeeney, and Alejandra Gimenez. Evaluating online nonprobability surveys. https://www.pewresearch.org/methods/2016/05/02/evaluating-online-nonprobability-surveys/, May 2016.

[KNN04]   Michael H. Kutner, Chris Nachtsheim, and John Neter. *Applied Linear Regression Models.* McGraw-Hill, 2004.

[LC20]   Sam Levine and Alvin Chang. Democrats took a risk to push mail-in voting. it paid off. Internet, 2020. [Online; accessed 25-March-2024].

[Lei22]   David Leip. Dave Leip's Atlas of U.S. Presidential Elections. Internet, 2022.

[Ley96]   Eduardo Ley. On the particular distribution of the U.S. stock indexes' digits. *The American Statistician*, 50(4):311–313, November 1996.

[Lij99]   Arend Lijphart. *Patterns of Democracy: Government Forms and Performance in Thirty-six Countries*. Yale University Press, 1999.

[LR02]   Roderick J. A. Little and Donald B. Rubin. *Statistical Analysis with Missing Data*. Wiley, 2nd edition, 2002.

[Lus19]   Ellen Lust. *The Middle East*. CQ Press, 15th edition, 2019.

[Mae17]   Maeil Broadcasting Network. [19th presidential election] Moon Jae-in solo 39.2%... Ahn Cheol-soo, Hong Joon-pyo's second-place fight. Internet, 2 May 2017. http://www.mbn.co.kr/pages/vod/programView.mbn?bcastSeqNo=1154251.

[MAH14]   Roderick McInnes, Steven Ayres, and Oliver Hawkins. Scottish Independence Referendum 2014: Analysis of results; Research Paper 14/50. Internet, 30 September 2014. https://researchbriefings.files.parliament.uk/documents/RP14-50/RP14-50.pdf.

[Meb10]   Walter R. Mebane, Jr. Fraud in the 2009 Presidential Election in Iran? *Chance*, 23(1):6–15, March 2010.

[MJOS20]   Amy Mitchell, Mark Jurkowitz, J. Baxter Oliphant, and Elisa Shearer. Legitimacy of voting by mail politicized, leaving americans divided. Internet, 2020. [Online; accessed 25-March-2024].

[MK22]   Regina Mihindukulasuriya and Pia Krishnankutty. Who will succeed Gotabaya when he resigns? Game of Thrones begins to pick next Lanka President. Internet, 12 July 2022. https://theprint.in/theprint-essential/who-will-succeed-gotabaya-when-he-resigns-game-of-thrones-begins-to-pick-next-lanka-president/1035563/.

[MN89]   P. McCullagh and John A. Nelder. *Generalized Linear Models*. Monographs on Statistics and Applied Probability. Chapman & Hall/CRC, 2nd edition, 1989.

[MN00]   P. McCullagh and John A. Nelder. *Generalized Linear Models*. Chapman and Hall/CRC, 2000.

[Mon16]   Monmouth University. Clinton Leads by 6 Points. https://www.monmouth.edu/polling-institute/reports/monmouthpoll_us_110716/, 2016.

Bibliography 325

[Moo16] Peter Moore. Over-65s were more than twice as likely as under-25s to have voted to Leave the European Union. Internet, 27 June 2016. https://yougov.co.uk/topics/politics/articles-reports/2016/06/27/how-britain-voted.

[MS04] Walter R. Mebane, Jr. and Jasjeet S. Sekhon. Robust estimation and outlier detection for overdispersed multinomial models of count data. *American Journal of Political Science*, 48(2):392–411, April 2004.

[MT11] Henrik Madsen and Poul Thyregod. *Introduction to General and Generalized Linear Models*. Chapman & Hall/ CRC Press, 2011.

[Nat17] National Election Commission. Results of Presidential Elections.xls. Internet, 2017. https://www.nec.go.kr/engvote_2013/main/download.jsp?num=492&tb=ENG_NEWS.

[New81] Simon Newcomb. Note on the frequency of use of the different digits in natural numbers. *American Journal of Mathematics*, 4(1):39–40, 1881.

[New17] New York Times. Presidential election results: Donald J. Trump wins. https://www.nytimes.com/elections/2016/results/president, 2017.

[New19] News First. SLFP to support Gotabaya Rajapaksa. Internet, 9 October 2019. https://www.newsfirst.lk/2019/10/09/live-blog-slfp-to-support-slpp-presidential-candidate/.

[Nig11] Mark Nigrini. *Forensic Analytics: Methods and Techniques for Forensic Accounting Investigations*. Wiley-Corporate F&A, 2011.

[Nig12] Mark Nigrini. *Benford's Law: Applications for Forensic Accounting, Auditing, and Fraud Detection*. Wiley-Corporate F&A, 2012.

[NP33] Jerzy Neyman and Egon S. Pearson. On the problem of the most efficient tests of statistical hypotheses. *Philosophical Transactions of the Royal Society of London. Series A, Containing Papers of a Mathematical or Physical Character*, 231:289–337, 1933.

[NSCS18] Tyler Nelson, Joon Jin Song, Yoo-Mi Chin, and James D. Stamey. Bayesian correction for misclassification in multilevel count data models: An application to the impact of exposure to domestic violence on number of children. *Computational and Mathematical Methods in Medicine*, 2018, 2018. https://www.hindawi.com/journals/cmmm/2018/3212351/.

[NW72] John A. Nelder and Robert W. M. Wedderburn. Generalized linear models. *Journal of the Royal Statistical Society. Series A (General)*, 135(3):370–384, 1972.

[oHR16] European Court of Human Rights. Party for a democratic society (dtp) and others v. turkey - 3840/10, 3870/10, 3878/10 et al. Internet, 2016. https://www.eods.eu/elex/uploads/files/5c863bfe20ad6-Party%20for%20a%20Democratic%20Society%20(DTP)%20and%20Others%20v.%20Turkey.pdf.

[Pie14] Pierre-Simon, Marquis de Laplace. *Essai Philosophique sur les Probabilités*. Courcier, 1814.

[Put01] Robert Putnam. *Making Democracy Work*. Princeton University Press, 2001.

[Qui19] Quinnipiac Polls. Warren up, Harris down, but Biden Leads among U.S. Dems Quinnipiac University National Poll Finds; 14 Dems Have Less than 1 Percent. https://poll.qu.edu/images/polling/us/us08062019_ubrt73.pdf, 2019.

[R C18] R Core Team. *R: A Language and Environment for Statistical Computing, v.3.4.3*. R Foundation for Statistical Computing, Vienna, Austria, 2018.

[RC88] Timothy R. C. Read and Noel A. C. Cressie. *Goodness-of-Fit Statistics for Discrete Multivariate Data*. Springer-Verlag, 1988.

[RE23] Ravi Rannan-Eliya. AK Dissanayake consolidates lead in presidential election voting intent. Internet, 24 November 2023. https://ihp.lk/news/pres_doc/IHPPressRelease20231124.pdf.

[Ric07] John A. Rice. *Mathematical Statistics and Data Analysis*. Brooks & Cole, 3rd edition, 2007.

[Rob07] Christian P. Robert. *The Bayesian Choice*. Springer Texts in Statistics. Springer Press, 2nd edition, 2007.

[R 23] R Core Team. *R: A Language and Environment for Statistical Computing*. R Foundation for Statistical Computing, Vienna, Austria, 2023.

[RT14] Kandethody M. Ramachandran and Chris P. Tsokos. *Mathematical Statistics with Applications in R*. Academic Press, 2nd edition, 2014.

[Sam89] S. W. R. de A. Samarasinghe. Sri Lanka's presidential elections. *Economic and Political Weekly*, 24(3):131–135, 1989.

[SBS15] SBS News. Sri Lanka's new president appoints PM. Internet, 10 January 2015. https://www.sbs.com.au/news/sri-lanka-s-new-president-appoints-pm.

[Sco13] Scottish Parliament. Scottish Independence Referendum Act 2013. Internet, 17 December 2013. https://www.legislation.gov.uk/asp/2013/14/introduction.

[Sha03] Amita Shastri. Sri Lanka in 2002: Turning the Corner? *Asian Survey*, 43(1):215–221, 2003.

[Sha07] Jun Shao. *Mathematical Statistics*. Springer Texts in Statistics. Springer, 2007.

[Sin16] Matt Singh. Polls apart. https://www.ncpolitics.uk/2016/03/new-polls-apart/, 29 March 2016.

[Ske48] J. G. Skellam. A probability distribution derived from the binomial distribution by regarding the probability of success as variable between the sets of trials. *Journal of the Royal Statistical Society: Series B (Methodological)*, 10(2):257–261, 1948.

[Smi02] Richard L. Smith. A statistical assessment of Buchanan's vote in Palm Beach County. *Statistical Science*, 17(4):441–457, 2002.

[Smi17] Martin R. Smith. Ternary: An R package for creating ternary plots. *Comprehensive R Archive Network*, 2017.

[Smi23] Martin R. Smith. Plottools: Add continuous legends to plots. *Comprehensive R Archive Network*, 2023.

[Sri24] Sri Lanka Brief. Election violence. Internet, 2024. "[Online; accessed 27-March-2024]".

[Sup23] Supreme Election Council. Cumhurbaşkanı Seçimi ve 28. Dönem Milletvekili Genel Seçimi. Internet, 2023. https://www.ysk.gov.tr/tr/14-mayis-2023-secimleri/82491.

[Sur14] Survation. 24 Hour Scottish Referendum Poll. https://survation.com/wp-content/uploads/2014/09/24-hour-scottish-referendum-poll.pdf, 2014.

[SYR06] Clémence Scalbert-Yücel and Marie Le Ray. Knowledge, ideology and power. deconstructing Kurdish studies. *Ecological Monographs*, 5:685–701, 2006.

[Tam14] TamilNet. Maithiripala sirisena of slfp emerges as common opposition candidate contesting Rajapaksa, Nov 2014.

[Tar16] Brian Tarran. The Economy: A Brexit vote winner? *Significance*, 13(2):6–7, 2016.

[US 04] US Department of State. Background Notes on Countries of the World: Democratic Socialist Republic of Sri Lanka, January 2004.

[US 19] US Census Bureau. Female persons, percent. https://www.census.gov/quickfacts/fact/table/US/SEX255218, 2019.

[Uya10a] Jayadeva Uyangoda. Sri Lanka After the Presidential Election. *Economic and Political Weekly*, 45(6):12–13, 2010.

[Uya10b] Jayadeva Uyangoda. Sri Lanka in 2009: From Civil War to Political Uncertainties. *Asian Survey*, 50(1):104–111, 2010.

[VR03] W. N. Venables and Brian D. Ripley. *Modern Applied Statistics with S-Plus*. Springer, 2003.

[VRG16] Pieter Vermeesch, Alberto Resentini, and Eduardo Garzanti. An R package for statistical provenance analysis. *Sedimentary Geology*, 336:14–25, 2016.

[Vyr09] Vyriausioji Rinkimų Komisija. 2009 m. gegužės 17 d. Respublikos Prezidento rinkimai. Internet, 2009. https://www.vrk.lt/statini ai/puslapiai/2009_prezidento_rinkimai/output_lt/rezultat ai_vienmand_apygardose/rezultatai_vienmand_apygardose1tu ras.html.

[Wan99] Leonard Wantchekon. On the nature of first democratic elections. *The Journal of Conflict Resolution*, 43(2):245–258, 1999.

[Wat09] Human Rights Watch. Turkey: Kurdish party banned. Internet, 2009. https://www.hrw.org/news/2009/12/11/turkey-kurdish-party -banned.

[WDA14] Carolyn Warren, Kimberly Denley, and Emily Atchley. *Beginning Statistics*. Hawkes Learning Systems, 2nd edition, 2014.

[Whi80] Halbert White. A heteroskedasticity-consistent covariance matrix estimator and a direct test for heteroskedasticity. *Econometrica*, 48(4):817–838, May 1980.

[Wij09] Ranil Wijaypala. Tiger leader counts his final hours. Internet, 17 May 2009. http://archives.sundayobserver.lk/2009/05/17/se c04.asp.

[Wi19] Wikipedia contributors. Opinion polling for the United Kingdom European Union membership referendum— Wikipedia, The Free Encyclopedia. Internet, 11 November 2019. https://en.wikipedia.org /w/index.php?title=Opinion_polling_for_the_United_Kingdo m_European_Union_membership_referendum&oldid=925644094.

[Wi20a] Wikipedia contributors. 2008 United States presidential election in Oklahoma. Internet, 2020. https://en.wikipedia.org/w/index.p hp?title=2008_United_States_presidential_election_in_Okl ahoma&oldid=952116013.

Bibliography

[Wi20b] Wikipedia contributors. 2014 Scottish independence referendum. Internet, 2020. https://en.wikipedia.org/w/index.php?title=2014_Scottish_independence_referendum&oldid=947796451.

[Wi20c] Wikipedia contributors. 2016 United Kingdom European Union membership referendum. Internet, 2020. https://en.wikipedia.org/w/index.php?title=2016_United_Kingdom_European_Union_membership_referendum&oldid=953969194.

[Wi20d] Wikipedia contributors. 2016 United States presidential election — Wikipedia, The Free Encyclopedia. Internet, 4 February 2020. https://en.wikipedia.org/w/index.php?title=2016_United_States_presidential_election&oldid=939055051.

[Wi20e] Wikipedia contributors. 2017 South Korean presidential election. Internet, 2020. https://en.wikipedia.org/w/index.php?title=2017_South_Korean_presidential_election&oldid=953913768.

[Wi24a] Wikipedia contributors. 2016 United States presidential election. https://en.wikipedia.org/w/index.php?title=2016_United_States_presidential_election&oldid=1214405867, 2024. [Online; accessed 20-March-2024].

[Wi24b] Wikipedia contributors. 2020 United States presidential election. https://en.wikipedia.org/w/index.php?title=2020_United_States_presidential_election&oldid=1213646580, 2024. [Online; accessed 20-March-2024].

[Wi24c] Wikipedia contributors. Opinion polling for the 2023 turkish parliamentary election. Internet, 2024. [Online; accessed 5-March-2024].

[Wil27] Edwin B. Wilson. Probable Inference, the Law of Succession, and Statistical Inference. *Journal of the American Statistical Association*, 22(158):209–212, 1927.

[WMS08] Dennis Wackerly, William Mendenhall, and Richard L. Scheaffer. *Mathematical Statistics with Applications*. Thomson Brooks/Cole, 7th edition, 2008.

[Woo06] Simon Wood. *Generalized Additive Models: An Introduction with R*. Chapman & Hall/ CRC Press, 2006.

[Wor19] World Bank. Population, female (% of total population). https://data.worldbank.org/indicator/SP.POP.TOTL.FE.ZS, 2019.

[WSS+01] Jonathan N. Wand, Kenneth W. Shotts, Jasjeet S. Sekhon, Walter R. Mebane, Jr., Michael C. Herron, and Henry E. Brady. The butterfly did it: The aberrant vote for Buchanan in Palm Beach County, Florida. *American Political Science Review*, 95(4):793–810, December 2001.

[Wu01] Joseph Wu. DPP shuffling toward the center. Internet, 30 October 2001. https://www.taipeitimes.com/News/editorials/archives/2001/10/30/109332.

[WW97] Peter H. Westfall and Russell D. Wolfinger. Multiple tests with discrete distributions. *The American Statistician*, 51(1):3–8, 1997.

[Yì0] Měng Yàn. Gbagbo, Ouattara to enter 2nd round of presidential election in Cote d'Ivoire. Internet, 4 November 2010. http://english.people.com.cn/90001/90777/90855/7188795.html.

[Yee98] Thomas W. Yee. A new technique for maximum-likelihood canonical gaussian ordination. *Ecological Monographs*, 74:685–701, 1998.

[Yee10] Thomas W. Yee. The vgam package for categorical data analysis. *Journal of Statistical Software*, 32:1–34, 2010.

[Yee15] Thomas W. Yee. *Vector Generalized Linear and Additive Models: With an Implementation in R*. Springer, 2015.

[Yeo17] Yeouido Institute. Public opinion poll at Yeouido Institute..Moon 39.4 Hong 24.9 Ahn 20.1. Internet, 3 May 2017. http://www.donga.com/ISSUE/2017president/News?gid=84188748&date=20170503.

[You79] Mary Sue Younger. *A Handbook for Linear Regression*. Duxbury, 1979.

[You16] YouGov. YouGov Survey Results. Internet, 23 June 2016. https://d25d2506sfb94s.cloudfront.net/cumulus_uploads/document/640yx5m0rx/On_the_Day_FINAL_poll_forwebsite.pdf.

[YW96] Thomas W. Yee and C. J. Wild. Vector generalized additive models. *Journal of the Royal Statistical Society. Series B (Methodological)*, 58(3):481–493, 1996.

Index

attach, **130**, 140
cbind, **216**, 263
chisq.test, 199
colnames, 160
confint, 263
dGBenford, 185
dirichletPrior, 163
family, **215**
getInitDigitDistribution, 199
glm, **215**, 261
head, 159
ifelse, 144
link, **215**
lm, 136, 211, 213
mean, 160
pbeta, 59
predict, 139
qbeta, 59
qchisq, 217
qnorm, 87
quantile, 33
quasibinomial, 217
rbinom, 33, 127
seq, 77
summary, 211, 218
vglm, 218, 261, 263

Afghanistan, 170, 186, 190
Agresti-Coull estimator, **26**
Archimedes, 181
Australian ballot, 241, 249

ballot box stuffing, 171, 207
ballot curing, 243
Bayes' Law, 158
Bayes, Rev. Thomas, 103
Bayesian analysis, 76, 100, 142

Advantages, disadvantages, **101**
Bayes' Law, 55, **103**, 104
Conjugate Prior, **108**
conjugate prior, 158, 310
credible interval, 55, 101, 158
likelihood function, 57, 105
posterior distribution, 57, 105, 107
prior distribution, 57, 58, 105, 159
Benford test, 172, **175**, 176, **199**, 249
Benford, Frank, 173
Beta, 57
Beta distribution, 158, **309**
Betabinomial distribution, 218, 261
bias, 7, 17, **25**, 26, 71, 77, 279
  missing data bias, 11
  nonresponse bias, 11
  participation bias, 11
  response bias, 10
  selection bias, 10
Binomial distribution, 24, 106, 214, 215, 227, 279, **297**
Binomial procedure, 130
Binomial test, **303**
blackout period, 122, 131
Bolivia, 161
Bonferroni adjustment, 155, 305
Brexit, 229

Côte d'Ivoire, 187, 190, 201
Central Limit Theorem, 92, 100
chi-square test, 177
conditional probability, 103
confidence interval, 50, **53**, 64, 85
  comparing, 127
  region, 156, 164

331

confounding variable, 239
consistency, **28**
count fraud, 172, 190
coverage, 64, 92, 125, **127**, **149**
cross-tab, 8, 87
CVR file, **242**

demographics, 244
design matrix, 224
differential invalidation, **204**, 243, 259
Dirichlet distribution, 158, **311**

Electoral College (US), 243, 277
electoral forensics, **172**
estimator comparison, **25**
Eudoxus of Cnidus, 181
European Union, 229
expected value, **39**

first past the post, 121
free and fair, **172**, 203, **210**
frequentist analysis, **60**, 100, 142, 189, 250

The Gambia, 151
gender gap, 89
generalized Benford, **180**, 182, 183
generalized Benford test, **180**, 183, 197
    likelihood simulation, **184**
    multinomial averaging, **188**
generalized linear model, *see* regression, GLM
Ghana, 176, 203

homoskedasticity, 136, 211, 212, 260
house effects, 73, 76, **79**
Hypergeometric distribution, 35
hypothesis, 250

invalidation plot, 207, **207**

Kurdish independence, 98

Law of Anomalous Numbers, 175
Law of Large Numbers, 100

leading-digit function, 195
likelihood, 295
likelihood function, 184
linear predictor, **215**
linear regression, 135
    ordinary, **135**, 146
    weighted, **136**, 139, 141
link function, 215, 216
Lithuania, 186, 190
logit function, 216

margin of error, **45**
maximum likelihood, 217
maximum quasi-likelihood, 217
mean, *see* expected value
mean squared error, 17, **29**, 30, 80
missing data, 280
model selection, 219
MSE, *see* mean squared error
Multinomial distribution, 158, **308**
multiple comparisons problem, **153**, 304

Newcomb, Simon, 173
non-probabilistic sampling, 8

ordinary least squares, *see* regression, OLS
outlier, 252
overdispersion, **217**, 219, 258, 260
oversample, 71
overvote, 243

point estimate, 45
polling
    assumptions, 34
    averaging, 124, 128
    combining, 123–142, 236
    legitimate, 22
    modality, 232, 237
    weighting, 237
population
    sampled, 279
    target, 279
population of interest, 5
precision, 28, 127

Index 333

prediction, 134
probability density function, 106
probability distribution
　Benford, **195**
　integer Benford, 195
　log-uniform, **194**
probability function, 296
purposive sampling, 9

R Statistical Environment, **xvi**, 282
random variable, 39, 295
regression
　Betabinomial, 218, 219, **261**
　Binomial, 214, 219, 260
　GLM, **214**, 260
　OLS, 210, 219, 260
　VGLM, 218, 260
　WLS, 212, 219, 260
Republic of Korea, 121

sample, 5, 22, 24
sample size, 24
sampled population, 5
sampling frame, 24
sampling scheme, **24**
Scotland, 21
significance
　practical, 77
　practical vs. statistical, **212**
simple random sampling, 7, 14, **24**, 81, 279
　bias, 25
　mean squared error, 29
　variance, **28**, 32
　without replacement, **35**
Sri Lanka, 257

SRS, *see* simple random samplnig
statistical thinking, 250
stats is evidence, 179
stratified sampling, 8, 69, 91, 94, 237
　bias, **74**, 77
　mean squared error, **79**
support, 235

Türkiye, 98
target population, 5
term limit, 2
Two Policemen and a Drunk, 181

undervote, 243
Uniform distribution, 106
United Kingdom, 21, 229
　St. Ives, 312
United States, 68
　Florida, 241
　Galesburg, 311
　Oklahoma, 185, 189

variability, 33, 279
variance, 7, 17, **40**
vector GLM, *see* regression, VGLM
voter, 233
　likely, 233
　undecided, 234

Wald procedure, 130, 133
weighted least squares, *see* regression, WLS
weighting data, 72, 237
weighting function, **132**, 133, 139, 141

For Product Safety Concerns and Information please contact our
EU representative GPSR@taylorandfrancis.com Taylor & Francis
Verlag GmbH, Kaufingerstraße 24, 80331 München, Germany